COPING WITH NATURAL ENVIRONMENTS

Glen A. Marotz
Robert W. McColl
University of Kansas

KENDALL/HUNT
PUBLISHING COMPANY
Dubuque, Iowa

Copyright © 1982 by Kendall/Hunt Publishing Company

Library of Congress Catalog Card Number: 81-85722

ISBN 0-8403-2651-3

All rights reserved. No part of this publication may be reproduced, stored in a retrieval system, or transmitted, in any form or by any means, electronic, mechanical, photocopying, recording, or otherwise, without the prior written permission of the copyright owner.

Printed in the United States of America

B 402651 01

Contents

Figures vii
Tables ix
Preface xi

I. The Relationship Between Humans and Natural Environment 1
  Viewpoints 1
    Environmental Determinism 1
    Possibilism 2
    Probabilism 3
    Ecological Anthropology 3
    Human Determinism 4
    Which Is the Correct Viewpoint? 6
  Terms 7
  References 8

II. Cyclic Phenomena 9
  Natural Cycles 10
    Agricultural 10
    Astronomical 13
      Solar 14
      Lunar 15
    Atmospheric 15
    Biological 18
    Cultural 21
    Geological 21
  Concluding Thoughts 22
  Terms 24
  References 26

III. Environment and *Homo Sapiens* 27
  Physiological Stresses 27
    Water and Temperature Stresses 28
    Biochemical Constraints 29

Diet	32
Sources of Dietary Elements	33
Historical Changes	34
Homeostatic Mechanisms	36
Thermal Equilibrium	36
Hypothermia	40
Hyperthermia	41
Altitude Problems: Hypoxia	43
Some Morphological Adaptations to the Environment	44
Terms	46
References	48

IV. Interpreting Natural Environments   49

Biological Systems	49
The Triune Brain	49
Some Possibilities	51
Processing Information	52
Information and Decision-Making	52
Social Behavior	56
Territoriality	56
Space Classification	56
Terms	58
References	59

V. Natural Environments   61

Environmental Structure	61
Biome Structure and Function	61
Productivity	63
Biome Concept and Human Strategies	64
Major Biome Types and Energy Relationships	67
Biome Characteristics	73
Tropical Biomes	73
Temperate Biomes	77
Cold Biomes	82
Deserts	85
Terms	86
References	88

VI. Adaptive Responses to Biomes: Clothing and Housing   89

Clothing and Housing as Responses to Environment	89
Clothing	91
Housing	92

Clothing and Housing as Adjustments to Environments in Different Biomes	**95**
Tropical Rainforest	**95**
Monsoon Forest	**97**
Tropical Savanna	**98**
Deserts	**99**
Prairie/Steppe Biome	**100**
Temperate Evergreen Forests	**102**
Temperate Deciduous Forests	**103**
Temperate Rainforest	**105**
Boreal Forest	**105**
Tundra	**106**
References	**109**
VII. Closing Thoughts	**111**
Index	*113*

Figures

1.1.	Spaceship earth	5
2.1.	A simple vegetable garden crop calendar	12
2.2.	Seasons, daylength, earth-sun geometry	14
2.3.	Phases of the moon and tide-producing gravitational effects	16
2.4.	An edge view of the atmosphere	17
2.5a.	Preferred hours of study among students	19
b.	Circadian rhythms of one isolated person	20
3.1.	An ecotonal area between grasses typical of the Great Plains and coniferous trees typical of the lower elevations of the Rocky Mountains	28
3.2.	Energy exchanges between humans and the environment	38
3.3.	Wind chill	41
3.4.	Water needs under various temperature conditions	42
3.5.	Surface area to body mass ratio among foxes	45
4.1.	A highly simplified representation of the reptilian complex, limbic system, and neocortex in the human brain	50
4.2.	Pyramid of human needs	52
4.3.	A simplified model of the decision-making process	53
4.4.	Sensory areas of the body	54
5.1.	Interaction of physical and biological factors in natural environments	62
5.2.	Structure, function, energy flows and nutrient cycling in biomes	63
5.3.	Generalized estimate of biomass productivity	65
5.4.	Stresses and adjustments in hypothetical natural (a) and energy subsidized (b) environments	66
5.5.	Generalized model assemblages of biome structure	68
5.6.	Simplified position and environmental conditions of model biomes arranged on a hypothetical continent	70
5.7.	Some changes in biome characteristics with changing environmental conditions	72
5.8.	Typical rainforest overview	74
5.9.	Slash and burn agriculture	75
5.10.	Rice fields in a monsoon forest biome	76
5.11.	Temperate rainforest biome	78
5.12.	Temperate evergreen woodland	79

5.13.	Temperate deciduous biome	80
5.14.	Prairie-steppe environment	82
5.15.	Boreal forest overview	83
5.16.	Tundra landscape	84
5.17.	Desert landscape	85
6.1.	Factors important for clothing and housing design	90
6.2.	Tropical rainforest clothing	96
6.3.	Monsoon housing	97
6.4.	Bedouin tent	98
6.5.	Caveform housing	99
6.6.	Exterior and interior conditions in an adobe dwelling	100
6.7.	Street scence in a desert biome	101
6.8.	Sod house on the Great Plains	102
6.9.	Modern earth-sheltered house	103
6.10.	Spanish ranchhouse fountain and courtyard	104
6.11.	New Orleans housing	105
6.12.	Boreal Forest housing	106
6.13.	Exterior and interior conditions for a snow cave	107
6.14.	Government subsidized housing on the arctic tundra	107
6.15.	Eskimo outerwear	108

Tables

2.1	A Brief Selection of Known and Speculative Cyclical Phenomena	11
3.1	Nutrients	30
3.2	Selected Brain-produced Hormones	31
3.3	Metabolic Requirements for Life Support of One Man Per Day Required Supplies	33
3.4	Generalized Time Scale for Selected Events That Improved Carrying Capacity	35
3.5	Approximate Number of Calories Produced (M) for Different Activities by an Adult of $1.7M^2$	37
3.6	An Example of Heat Loss to the Environment	39
3.7	Summary of Human Responses to Thermal Stress	39
3.8	Selected Effects and Responses to Hypoxia	43
5.1	Characteristics of Selected Biomes	70
6.1	Surface Area Percentages of the Human Body	92

Preface

The purpose of this book is to provide the reader with a brief introduction to relationships between humans and physical environmental factors which most directly affect us. It is not a textbook or a laboratory manual; neither is it an essay promoting any particular point of view. What it is is an attempt to draw together material from the fields of physiology, physical geography, anthropology, cultural geography and ecology, and to use the material to provide a rudimentary understanding of the adaptations and adjustments that the human species must, and has, made in order to cope with the range of earth's physical environments. We hope that our approach will help readers with little exposure to the topic to have a better understanding of the interplay between human physiological characteristics and the nature of the earth's biomes.

We have tried to remain objective as well as readable and informative. The complexity and comprehensive nature of the subject matter meant that assiduous summarization of details and much data was necessary. Where appropriate, tables, diagrams and references are provided to help explain what may seem controversial or unfamiliar ideas.

Structure

The book begins with an overview of the major philosophies of man/land relationships. We then proceed to a review of current information on the physiological and morphological characteristics of the body, and how they may determine reactions to various environmental factors. Emphasis is placed on the fragility of the body in terms of its temperature and chemical requirements. These human characteristics provide a base for examination of stresses placed on our systems in different environments. We next look at the interplay between environmental stresses and some of the body's natural homeostatic mechanisms for dealing with them.

Discussion of human physiological characteristics and environmental stress is followed by an introduction to the biome concept. Biomes are presented as energy-balanced systems that place certain natural limits on their use, limits that can only be overcome by adding energy (such as coal, electricity or oil) from some non-natural or non-local source. Man is viewed as a natural and integral part of many biomes, but each biome nevertheless presents constraints to which we must adjust and adapt. Biome richness and diversity and the implications for human use are discussed. Some are limited and must either be avoided or modified by modern technology if we are to function adequately.

Man is also seen as a unique feature of any natural environment. His actions are considered neither inherently bad nor good in such a situation. We discuss the implications that technology has for offsetting stressful conditions, and focus on the role that imported energy has in enhancing successful coping strategies. With an appreciation of the human body and the natural environment, it should be possible for the reader to reach his own conclusions as to whether we are constrained by environmental conditions or whether we can modify and change natural systems at will and at little or no cost to ourselves or the environment.

Additional information is contained in the references listed at the end of each chapter. Also, we have provided a list of the key terms used in each chapter to facilitate understanding and review. The references, term summaries and organization permit the book to serve as a reader in introductory college classes. We are using it in this fashion in two courses within the Geography-Meteorology Department at the University of Kansas: *Geography of Human Survival*, a freshman-sophomore class that has grown from an enrollment of twenty to about 300 students per semester, and *Environment and Man*, an introductory course offered for junior-senior credit to students from a wide variety of disciplines but with no prior background in the subjects treated in the following pages. *Environment and Man* has an enrollment of about 150 students per semester.

We are convinced that much of the success of both courses lies with the interest in man/land relationships and our use of information from a wide variety of disciplines. This approach also means that the book is useful in other contexts, such as architecture, urban planning, and environmental studies, as our colleagues have discovered. It can serve particularly well as a reader that elicits discussion in small groups or seminars.

We wish to thank Ms. Beverly Koerner for typing each of the endless "final draft" manuscripts. Any errors in the text are solely our responsibility.

The Relationship Between Humans and Natural Environments

The relationship between humans and their physical environment has been a matter of speculation, study, and even religion since the beginning of recorded history. Environmental threats, such as cold, heat, flood and drought, have always been a part of human existence. Water has been honored as a god. Its plants (the papyrus, bullrushes, lily and lotus) and animals (the crocodile, ibis and fish) have become major religious and cultural symbols to many peoples of areas where rivers and flooding are common (Egypt, Southeast Asia, India, China, the Indus Valley). In desert, grassland, and forest environments, the sun, fire and wind have dominated the symbols of indigenous cultures expressing an awareness of factors that confronted humans in these environments.

Seasons of cold and heat, light and dark, food and famine, all presented humans with a regularity of activity and passivity, and substantially influenced adaptation and adjustment strategies. As humans became more technologically advanced, they often were able to modify, even overcome, some natural environmental threats to life and comfort. These abilities brought about new views of the earth, new philosophical and religious contexts. The question of the relationship between humans and the natural environment remains a dominant theme in all science, technology and philosophy today, and will remain so in the future as well. Let us begin our survey of human/environmental relationships by outlining some of the principle philosophical views.

Viewpoints

Environmental Determinism

In this view, humans are dominated by the environment which predicates almost total subservience or vulnerability to natural laws. The idea is an old one originating with the Greeks and is traceable through the nineteenth and early twentieth centuries in the writings of Huntington, Ratzel, Semple and others. The philosophy is expressed in such phrases as "the laws of nature," the "natural limits of . . . ," the "balance of nature," or terms such as nature "controls," "determines," and similar words. It is a view that was extended to explain social and political events, such as the rise of Great Britain and Japan as major world powers due to the "influence of their island locations." It purports to explain the rise of monotheism in the Middle East (both Christian and Moslem) as a manifestation of the impact of the desert, of survival based upon individual achievement, and of the dominance of singular elements, such as the sun or water versus the multitude of elements found in the tropics.

At one time, most scholars accepted determinism as a natural consequence of Darwin's theories on evolution. Some of our contemporaries still stress the view that nature has certain finite limits which cannot be exceeded without causing damage and ultimately affecting human survival. Thus, even contemporary man, despite all of his science and technologoy, is "forced" by inherent biological limits to use certain behavioral and even physiological characteristics if the species is to successfully cope with altered environments.

Environmental determinism, whether carried to the extreme that views nature as dominant over human activity, or that the environment merely is the single most important element affecting man's successful occupation of the planet, places human beings in a very limited state of free will and permits little ingenuity. While this view may appeal to some poets like Frost or Whitman, or philosophers such as Lao Tzu, it is a view that is unsettling to many others who feel that we are somehow above, and better able to cope than other animals with any but the most dramatic and severe of natural forces.

Possibilism

Humans having little or no say in their own destiny was an unappealing idea to many. To provide a more emotionally, as well as intellectually, acceptable point of view, the idea emerged that while the environment clearly sets physical limits as to what we can and cannot do, it still offers a range of possible uses; we are essentially free to choose from among them, but culture and history have an effect on what is selected. Franz Boaz first stated the concept in the late nineteenth century as a direct refutation of determinism.

This viewpoint also permits expansion of the range of possible uses of an environment under even extreme conditions through technology. For example, swimming pools and palm trees were included in special buildings during the construction of the Alaskan Pipeline. Thus, the environment inside the buildings was more like Florida than Alaska. However possibilists would say that the natural environment at some point sets limits which cannot be exceeded without increasing the threat of disaster, or the end of human occupance of that environment.

Probabilism

While possibilism met the philosophical need to feel more like an equal in our relationship with nature, it did not, and does not, deal with the fact that few people see the same range of possibilities in the same environment. Early American Indians, later the Spanish padres, and finally the American colonists, lived in the desert southwest and southern California. Each group used the environment differently; each perceived a different use or limits of the same environment. Native Americans were largely hunters and gatherers; the Spaniards, more ranchers and cattlemen; the American colonists, farmers, miners, merchants and city builders. It is also clear that if you placed a Chinese, a Frenchman, and an Englishman in the original swamps and bayous of the Mississippi delta, each would see a different set of possibilities and each would react differently and create a different kind of cultural landscape. Why?

Alexander Goldenweiser writing in the 1930's was among the first to argue that, while the environment offers a range of possible human uses and we are free to choose among them, such freedom is largely hypothetical. The environment was a static element in the man/land mix. Culture, traditions, past experience, were all dynamic and favored conceptions of how an environment can be used. We all know of examples of land that has been reclaimed from tidal marshes, swamps or garbage dumps and that today houses entire cities. Thus, the Probabilist contends that while the natural environment may set physical limits to use of the earth's surface, actual uses are more related to cultural background and preconceptions. Culture thus moved to a more dominant position in the man/land interface.

Ecological Anthropology

Application of these viewpoints to the study of human/environmental relationships led to problems. Determinism, for example, could not account for the vastly different native cultures that occupied the Mediterranean basin and similar environments along the temperate west coast of continents. Julian Steward is credited with proposing a different view, cultural ecology now evolved to *ecological anthropology,* that many people feel accounts for such disparities.

The idea is elegantly simple. Neither environment nor culture serves as the control over human/environmental interaction. Rather, man/environmental relationships are governed by the need to use appropriate strategies for resource utilization in a particular environment. Physiological, cultural, and behavioral adjustments and adaptations needed to insure efficient resource utilization can be studied and compared using this approach, avoiding a need to justify either man or the environment as dominant.

What framework should serve as a uniform structure in order to decipher man/land interaction? If we assume that humans are subject to the same natural laws as other species, then Steward, Rapoport, Hardin, and Odum, among others, advocate the usage of terminology from ecology and ecosystemic theory as appropriate. The emphasis is biological, and most of the concepts involve energy flow and material cycling through units called biomes.

Steward argued that the interaction between humans and environment that results in appropriate resource partitioning seems to follow typical patterns no matter where the people or environment are located. The method breaks down as increasing technological advances allow humans to modify the environment itself, which leads to totally new ways of resource utilization that often include areas outside of the original environment. Humans may no longer utilize resources in a fashion entirely in tune with the local environment as technological capabilities grow, but the human/land relationship may be consistent with the range of environments used by a particular cultural group. An example would be irrigated maize farming in western Kansas. Such an agricultural activity is not consistent with the natural resources of the region, but is appropriate for such an area within the technology of contemporary North American populations.

Human Determinism

Our ability to live outside of natural constraints led to the view that, with technology and human ingenuity, there is no real or perceived environmental limit that cannot be overcome. We can aircondition to offset extreme heat, heat to offset cold. Deserts can be irrigated, swamps drained, mountains removed or tunneled if they pose barriers to travel. Science and technology are seen as the solution to environmental problems. Spectacular feats like the landings on the moon and the photos of earth obtained on the return reinforced this view in the minds of many Americans during the late sixties (Fig. 1.1).

Of course without the environment, there would be no need for technology to overcome threats and barriers to human survival. Thus, in a sense we have come full circle and the question of whether the environment determines human activity or human activity can overcome the environment is tautological. In fact, as we devise means to overcome real and perceived environmental barriers to human activity, we often create new, manmade environmental problems. For example, if the heating system in an arctic settlement should fail, what happens to life? While we can move mountains to build highways and canals, what do we do with the debris? We can use various fuels to provide rapid transportation but what happens to the quality of the air we must breathe? Obviously, these are questions of importance to all who are concerned with new problems technology creates as we spread our influence cross the face of the landscape.

The man/land relationship is also important to religion and self-esteem. Did, or did not, God give us dominion over the world? The problem remains one of the actual and the perceived relations between human physiological limits, and the natural limits created by the physical environment.

Figure 1.1. Spaceship earth photographed during the Apollo series of moon probes. James Lovell upon seeing this view of earth remarked, "We do not realize what we have on earth until we leave it." (Courtesy, NASA.)

Which Is the Correct Viewpoint?

Obviously each view is correct in some respect. Clearly, the natural environment does place limits on what we can and cannot do for physical survival and to populate various areas of the globe. The more extreme the environment—cold, heat, aridity or humidity—the more limiting it is. It also is clear that how an environment is used is often a reflection of how its assets and limits are perceived. Early American colonists avoided the plains because it was considered the Great American Desert. Today, it is the bread basket for many nations.

It also is evident that if we want to spend the money and material on completely modifying or remaking the environment, we can do this as well, but often at extreme cost. Finally, there is the fact that nature is more complex and interrelated (*symbiotic*) than has been man's ability to modify it. Too often human modification of a natural environment has changed the environment and created entirely new challenges to human ingenuity to overcome new problems and threats to human survival. The challenge of living in cities of twenty and fifty million—unable to produce their own food, burying themselves in garbage, unable to provide jobs, transportation and breatheable air—is a challenge of the present as well as the future.

We turn now to the question of just how dependent we are upon natural elements. In order to answer this question, we must understand the basic physiological and biological limits of *Homo sapiens,* as well the environmental factors that create various degrees of life threatening stress.

Terms

Ecological Anthropology A concept that places resource utilization by humans as the central focus of man/land studies. Terminology and approach are taken from ecological theory.

Environmental Determinism. A belief or philosophical view that emphasizes the physical (natural) environment as the cause of events and human behavior.

Human Determinism. This is the view, most strongly represented by Huxley, that man can remake and modify his environment to suit his own needs or desires.

Probabilism. The philosophical position that says that while an environment may offer a range of possible human uses, it is the background or culture of people that makes some of these choices or uses more likely than others.

Possibilism. This view of the man/land relationship is that all environments offer a range of possible uses by man and man is free to choose among these.

Symbiosis. A relationship between two or more organisms that may be of mutual benefit, but is always of benefit to one of the organisms.

References

Bates, M. *The Forest and the Sea.* New York: Vintage Books, 1953.

Boaz, F. "The Limitations of the Comparative Method of Anthropology." *Science,* 4:901, 1896.

Goldenweiser, A. *Anthropology.* New York: Crofts and Co., 1937.

Huntington, E. *Civilization and Climate.* New Haven: Yale University Press, 1915.

Rappoport, R. "The Flow of Energy in an Agricultural Society." *Scientific American,* 224:116, 1971.

Steward, J. *Theory of Cultural Change.* Urbana: University of Illinois Press, 1955.

II
Cyclic Phenomena

In all societies and cultures, there is a recognition of a cyclical or rhythmic nature to life. This recognition ranges from an awareness of the obvious and prominent physical *cycles* associated with lunar phases and the changing seasons of the year, to those which are biological or behavioral, like daily variations in body temperature and activity-rest periods. Among some religions, such as Buddhism and Hinduism, cycles of birth and re-birth are seen as central to all life.

Pattern recognition is the basis of all science or prediction. Cyclic phenomena follow repetitive patterns that allow us to predict and plan for future occurrences of the same event. With such knowledge, we are able not only to plan, but also to begin to modify either our environment or various physiological states, and, thus, to influence future outcomes. A familiarity with cyclic patterns puts life and behavior into a more predictable framework.

Cyclic phenomena are those that show a repetitive pattern. For example, seasons repeat themselves regularly and in the same sequence (summer, fall, winter, spring). Cyclical patterns are also marked by *periodicity,* or the regularity of the interval between occurrences. Thus, seasons are not only cyclical but they also are periodic. *Frequency* refers to how often a pattern normally repeats itself.

Some cycles are well defined, like the twenty-two year *sunspot* activity phenomenon, although the underlying causes for repetition may not be understood completely. Knowledge of cyclic events can be helpful in dealing with our surroundings even if causes for their occurrence are not known at the time of discovery. Indeed, our knowledge of new regularly recurring events is constantly growing, and adds to our ability to predict the workings of both natural systems and our own bodies and behavior. The adaptive role that cycles play is the subject of much debate among scientists. Most biologists would agree that cycles provide a framework for temporal organization of daily behavioral and physiological activities for a wide variety of organisms including man. The organizational frameworks are enough to make the study of natural rhythms important, but some biologists point to additional factors.

For example, sleep/awake cycles may have been early adaptive mechanisms with a number of advantages. *Nocturnal* (nighttime) and *diurnal* (daytime) species are active at different times and do not compete for the same food resources. If all species gathered similar foods day and night, there soon would be little left either for man or for the animals on which he fed. Obviously, sleep also decreases the need for food because activity and metabolism are reduced. This could be especially important in a harsh environment where great energy must be expended to gain even minimal food and where food choices are minimal. Hibernation and dormancy among animals certainly fulfills this role.

Sleep/awake patterns may also serve as protective devices; quiet sleep in a protected environment prevents predators from capturing prey that might well serve as a food source if both creatures were active at the same time, e.g., humans (diurnal) would have little chance against the jaguar (*Panthera onca*) which forages at night. Some biologists believe that this is one of the roles sleep played for humans as they developed as a species. Knowledge of the feeding and sleeping cycles of potential food sources can serve to enhance the food gathering process. This is important not only to humans but other species as well. For example, bees are active only during periods when flowers are open and nectar is available. This ability makes the forage process efficient since minimal energy is wasted in searching for food, a secondary benefit from the plant's perspective is that pollination will take place at precisely the right time. Yet another example of a gathering process based on cycles would be the scheduling of migratory habits among native Americans on the Great Plains to take advantage of seasonally available plant and animal resources.

Natural Cycles

A selection of some of the more widely known cyclic phenomena is shown on Table 2.1. The list is certainly not exhaustive and some of the items are controversial. This is particulary true for cycles that involve aspects of human behavior or concern long term events and, thus, limited data. The items serve as illustrations of the range of phenomena which at least some researchers see as having implications for a number of human activities.

Agricultural Cycles

The short term cycles listed on the table are familiar to most of us. For example, *growing seasons,* the time between first and last killing frost, define the limits of annual plant growth at any location. A knowledge of locally important growing seasons aids in determining what crops to plant and when they should be planted. Such knowledge can also be used to estimate the number of times the land can be cropped in one season.

Table 2.1. A Brief Selection of Known and Speculative Cyclical Phenomena

Cycle Name	Frequency	
	Short Term	Long Term
Agricultural	Growing Season Phenological Events Weather Patterns Disease and Insects	Climate

Astronomical	Solar Related
	  Day
	  Seasons
	  Sunspots
	Lunar Related
	  Tides
		Solar Radiation Variations

Atmospheric	Weather
	  Temperature
	  Moisture
	  Radiation
	  Wind
	  Pressure
		Climate
		  Wet Years
		  Dry Years

Biological	Circadian
	Weather Related
	  Mental Activity
	  Physiological Response
	  Disease
	  Accidents
	  Hibernation
	  Reproduction
		Life Stages

Cultural	Work/Play/Rest
	  Festivals
	  Religious Events
		Societal Evolution
		  Nation States

Geological	Glacial Epochs
	Landforming Processes
	  Aggradation
	  Degradation
	  Diastrophism
		Sea Level Changes
		Continental Drift

The length of the frost free period is highly variable from region to region and even within a given area. In the United States, growing seasons range from virtually year around (> 340 days) along the southern California coast, southern Texas and Florida, to about 120 days in northern North Dakota and Minnesota. Coastal locations generally have longer growing seasons because of the moderating influence of adjacent water bodies while continental, or inland locations, have shorter seasons. Topography, slope, and aspect also have an influence; higher elevations, and steep slopes that face away from the sun have more severe conditions and shorter growing seasons.

Phenological events are the cyclic life stages that *annual* plants progress through during part of or an entire growing season. The general stages are germination, emergence, growth, flowering, fruiting and death. Familiarity with the time of occurrence of each step can provide a predictable planting and harvesting calendar schedule called a *crop calendar* (Fig. 2.1). Crops may be planted or interplanted in such a way that an environment, which could support only a few people via unplanned agricultural practices, can potentially support more, or even produce a food surplus. This may be possible without adding any new technology, fertilizers or irrigation.

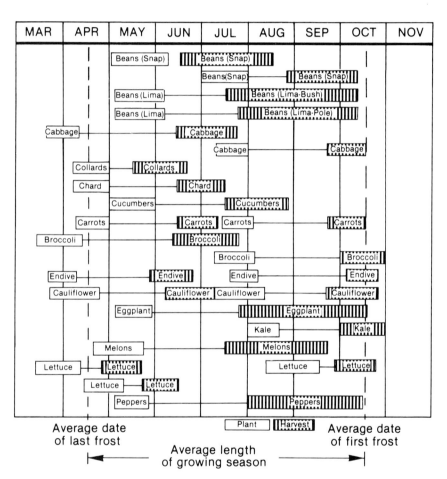

Figure 2.1. A simple vegetable garden crop calendar showing sowing, growth, and harvest dates for common crops; the chart is applicable to the central, eastern margin of the Great Plains. Note that some crops have resistance to frost after emergence and during harvest. (Data compiled by C. Marr, Kansas State Extension Service.)

Wild annuals, *biennials* and *perenials,* that may serve as food sources go through similar stages in a predictable time sequence. Scheduling of harvest activities is also possible, of course. Among hunter/gatherer populations, such harvest often meant or means significant movement across macrospace in order to arrive at the appropriate time.

Disease patterns are closely related to weather events and the availability of particular food sources. Some plant diseases, like wheat rust, are most destructive during warm, humid weather; others, such as potato blight, do damage following cool, rainy, continuously damp conditions. Since favorable weather for disease is likely to occur somewhere within the area where a desirable food source grows or is grown at about the same time every year, knowledge of typical disease occurrence patterns is essential if adjustments are to be made to any possible change in production levels.

All phases of plant development are susceptible to weather fluctuations. Different plants are affected differently by weather patterns; what may be good for one crop is not good for another. For example, farmers in a mid-latitude grassland might plant sorghum if a drier than normal series of weather conditions are expected in what is already a rather dry climate. Expectations of wet conditions might lead to more wheat planting than is typical. The crops grown by the farmer determine scheduling of machinery, cultivation, and harvest; regular cycles have to be altered to fit crop conditions.

The cyclical variations described above are short term, occurring over a season or less. The hunter-gatherer, rice farmer, and technologically-oriented mid-latitude farmers have adjusted to them and adopted strategies that allow existence under fluctuating weather conditions within a particular climate that stays essentially the same from year to year. There are long term climatic fluctuations, however, that cause more pronounced changes in agricultural coping strategies. For example, a warm series of years several hundred years ago permitted viticulture (grapes) in Greenland and England; grapes are no longer grown in either location because the average annual temperature has decreased. Droughts may be long term cyclical phenomena that have something to do with solar activity. Drier than normal conditions in the North American grasslands occurred in the 1890's, the second decade of the twentieth century, 1930's, 1950's and 1970's suggesting that there is a twenty year cycle that is related to the twenty-two year sunspot cycle; why this connection should exist is unclear. Nonetheless, cropping patterns, phenologic events, and production are all affected and must be adjusted to fit environmental conditions.

Astronomical Cycles

Astronomical cycles are the most obvious and precisely known of any that have an impact on human activity. They are still used as a basis for measuring time and denoting when cultural practices or events should occur.

Alternating day/night is one of the most important cycles because it affects the amount of sunlight available. Light in turn controls plant growth and thus crops and food. Seasonal transitions are only slightly less dramatic than the relatively short term day/night light variation. These too have long been noted and incorporated into human planning and adaptations. Naturally, the greater the seasonal differences, the more strongly have they been incorporated into cultural activities. Similarly, the more agrarian a society, the more important familiarity with seasonal shifts becomes.

Solar Cycles. Cycles that are well understood under this category include the most prominent of all, those associated with the sun and moon. Solar cycles are most evident in the regularity of the *seasons* produced by a combination of the earth's *rotation* around an inclined axis and *revolution* about the sun. The transition from one season to another is marked by the solstices and equinoxes. Every year on December 22 and June 21, the winter and summer *solstices* occur, a time when the sun in relation to the earth is at its most poleward position (overhead at either the Tropic of Cancer or the Tropic of Capricorn). Every March 21 and September 23, the sun's most direct rays are overhead at the equator and the days and nights are of equal length (*equinox*) (Fig. 2.2).

Times of seasonal change often are marked by religious holidays and various cultural rituals. Christmas is December 25th, more to mark the change from fall to winter than the birth of Christ; likewise the date chosen for Easter. Stonehenge, Ankor Wat, and many lesser known monuments and "holy" places are designed so that their features mark the passage of the seasons; people would know when to plant and harvest and could be reassured that life was following normal patterns. Disruptions, such as an eclipse, caused panic because such events were viewed as a potential end of the cycle, interpreted

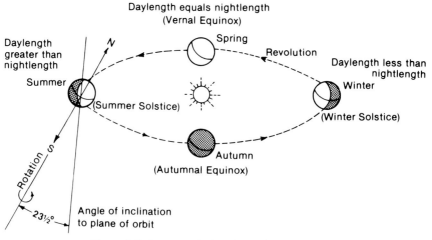

Figure 2.2. Seasons, daylength, earth-sun geometry

as the end of life. The ability to predict seasonal changes raised confidence that future food supplies would be adequate and that man had a certain primitive control over fate.

Lunar Cycles. Next to the sun and its cycles, few environmental occurrences are more notable than the regularity of lunar phases. For millenia, phases of the moon have been used to mark or note recurring events of life (Fig. 2.3). These periods are termed *synodic time.* They reoccur every 29.5 days, and are divided by the moon's phases—first quarter, second quarter, third quarter, or new, old, waxing or waning. No calendar is needed given the regularity of lunar cycles. Much of Asia still uses synodic time as the basis for calendars.

Certainly the most prominent lunar affect that influences life is the tidal fluctuation of the oceans and large bodies of water, such as the Great Lakes and the Caspian Sea. Tides follow regular and predictable patterns that not only affect biological life in the water and along the shore, but also human activities, such as fishing, that also depend upon such waters for food and livelihood. These influences clearly are gravitational and not associated with moonlight, but it is the light or phases of the moon that we note and relate to other events.

Atmospheric Cycles

Solar radiation variations are the most important item on our brief list because daily and seasonal cycles control the workings of the atmosphere and *biosphere,* the thin layer of the earth and atmosphere containing life processes (Fig. 2.4). Solar energy provides the driving force for the atmosphere, which responds by moving and by creating all of the phenomena that we observe in order to satisfy energy imbalances over the earth's surface. These motions and other phenomena result in weather and ultimately climate.

Climate may exhibit cyclical variations on very long or reasonably short time scales. Geological records show that there have been numerous alterations of warm and cold periods that have affected the presence and distribution of organisms all over the globe. Species of plants and animals now found only in warm, equatorial areas once lived in Antarctica where they could not survive today. During the last several million years, the climate of North America has alternated between present and very cold (glacial) conditions, marked by advance and retreat of large ice sheets across the continent. These alternating periods encompass almost the entire history of human existence. Our agriculture practices, life styles, technology, clothing and housing, all reflect familiarity and experience with only one part of a much larger cycle. What would a return to a very cold, glacial period imply to patterns of human distribution and activity? We do not know, although some researchers have speculated in detail. There are many proposed reasons for such climatic changes; some involve changes in insolation receipt resulting from progressive alterations in earth-sun geometry. We cannot tell whether or not they are

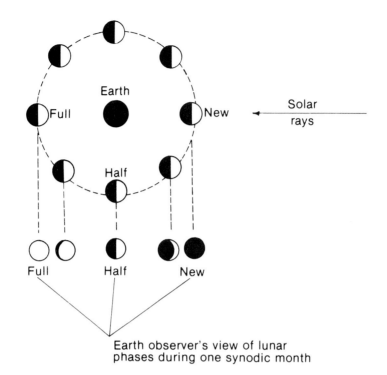
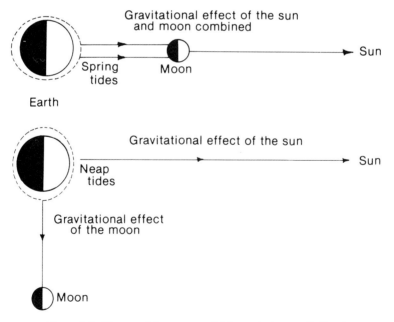

Figure 2.3. Phases of the moon and tide-producing gravitational effects

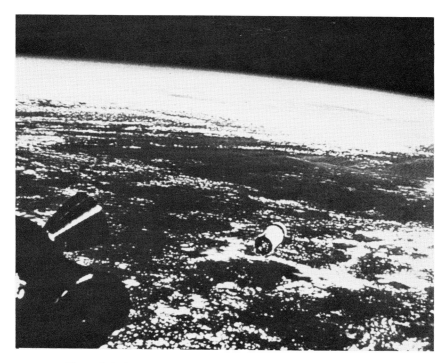

Figure 2.4. An edge view of the atmosphere. The biosphere occupies only a very small portion of the atmosphere, essentially the first two meters above and below the earth-atmosphere interface. (Courtesy, NASA.)

repetitive patterns because the data record is not very good. Some people have argued that cold conditions will return, probably faster than they might because of human burning of fossil fuels and resulting increases in the *Greenhouse effect*.

Seasonal and daily weather changes obviously occur within each climatic realm. In the mid-latitudes, winter is marked by alternating storminess, associated with large traveling low pressure weather systems, and brilliantly clear days when high pressure systems predominate. Summer usually has extended periods of warm, humid weather characterized by light southerly winds and thundershower activity. These patterns are repeated on a daily basis in both seasons. As we proceed north and south of the mid-latitudes, weather fluctuations decrease; daily and seasonal conditions become more like the principal climatic regime at a particular location.

Weather fluctuations appear to have an effect on certain behavior patterns and physiological functioning. Brezowsky has shown that reaction time for specific tasks is less under the high pressure, cool, mild, and dry conditions that prevail before and after passage of a mid-latitude cyclone. Lower pressures, warm, humid conditions associated in another part of the cyclone are associated with longer reaction times. He also notes that health clinics record a rise in asthmatic attacks, headaches, heart attacks and other ailments under weather conditions prevailing in the warm, humid section of cyclones. Other research has shown that behavioral changes may occur during certain weather events. A rise in crime statistics, auto accidents and irritability during periodic Santa Ana events (an episode of hot, dry winds across coastal cities in California) has been noted repeatedly. However, physical connections between the weather event and specific reasons for such changes have not been established.

Biological Cycles

There are a range of *endogenous* and *exogenous* cycles that affect human health, mood and behavior. The study of these biological rhythms (*biorhythms*) and their influences is called *chronobiology*.

Before looking at some of the more prominent biological cycles, we should note that virtually all are triggered or set in motion by some external stimuli called a *zeitgeber,* a German word meaning time-setter and referring to those exogeneous stimuli that act to synchronize our biological clocks and their rhythms to a common local time.

Biological cycles vary greatly from individual to individual. Without zeitgebers, cycles might vary from twenty-one to twenty-six hours. However, nature provides cues that help the body set its biological clocks to a kind of "standard" time. This latter time is keyed to seasons and latitude, which means that an individual adapted to one local environment would be out of synchronization or rhythm if suddenly moved to a drastically new location. However, local zeitgebers would soon adjust the individual's biological rhythms to local time in the new location. This ability of the body and its mechanisms to adapt in very short periods is one of the keys to understanding how we are able to migrate all over the planet—from the arctic to the tropics and from the mountains to depressions below sea level.

Light and temperature are the more important zeitgebers, or are at least the best understood at this time. In 1962, *Aschoff* proposed the following rule:

"Light accelerates the biological clocks of creatures normally active during the day, but delays the cycles of nocturnal creatures."

There is no more obvious zeitgeber than light. The length of light periods associated with the sun and its seasonal and daily cycles are well-known. It may be that myths and folk tales about the affects of seasonal changes upon human activity and behavior actually are based in part on changing light/dark regimes.

Some of the most pervasive cycles are those associated with the daily internal rhythms of our body. Some are more obvious, such as work and sleep, activity and resting; often they are associated with the daily light/dark pattern. For example, human body core temperatures average about 98.6°F (37°C), but they also follow a rhythmic daily cycle. Core temperature begins to rise during the morning hours; this rise is accompanied by increased metabolism and an increased heartbeat. From about noon to 2:00 p.m., there is a marked downward shift in all these rates, a period of decreased biological activity. All rates rise again in the afternoon, but not as high as in the early morning. All then fall until the nadir is reached normally at about 4:00 a.m. At this time, the body is most deeply at rest. To test the extent to which these rhythms are light related, volunteers have been tested in deep caves in the absence of natural light patterns. While their activity and temperature cycles varied from twenty-one to twenty-seven hour rhythms, thereby emphasizing the importance of exogenous zeitgebers, the rhythms were nevertheless regular.

Such research has shown that many of our internal biological cycles, blood chemistry, urine output, temperature, etc. vary on a regular basis regardless of the length of day or night (Fig. 2.5a, b). They are *endogenous*. The role of zeitgebers as local "fine-tuning" clocks thus takes on additional significance.

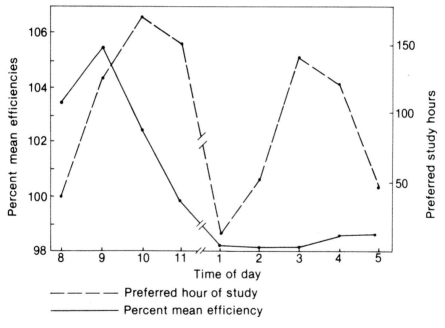

Figure 2.5a. Preferred hours of study (-) indicated by college students on a questionnaire versus actual performance (---) on five simple memory tests. Note that students did not study when their efficiency was greatest. (Adapted from Gates, A. 1916. Diurnal Variations in Memory and Association. Univ. Calif. Pub. in Psych. 1(5):323-344.)

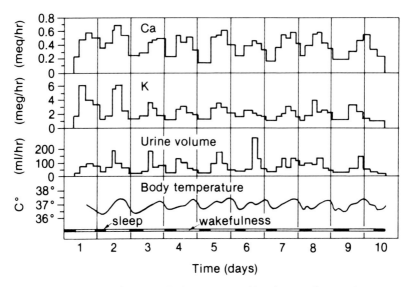

Figure 2.5b. Circadian rhythms of one subject in an underground bunker. Curves show excretory rhythms for calcium, potassium and urine. (Reprinted from J. Aschoff, 1965. Circadian Rhythms in Man. Science. 148:1427-1432. Copyright 1965 by the American Association for the Advancement of Science.)

Halberg's observation of endogenous rhythms led him to coin the term "circadian cycle" rather than adopt the term "diurnal" (day-night) cycle. *Circadian* is derived from the Latin roots *circa* (about) and *dia* (day) and refers to a basic twenty-four hour rhythm. It focuses on time and ignores the length of light and dark periods. This simple semantic distinction removes the subconscious fixation on light and darkness as causes of cycles and frees observers to look for patterns, regardless of the time of day or night they occurred. It also helps us to understand how such cycles can occur in people living under extremely different diurnal patterns—Eskimos versus inhabitants of the tropics, for example.

For many species, the diurnal as well as annual temperatures associated with seasonal change play key roles in affecting active and passive physiological responses. For example, the need to regulate body temperature to compensate for cold and warm periods elicits activity or passivity as a behavioral response. Temperature changes may also serve as an underlying longer range (seasonal, annual and perhaps longer than yearly) biological rhythms. Morbidity and mortality tables kept by life and health insurance companies clearly illustrate various repetitive patterns of sickness and death associated with particular times of the year. For example, respiratory diseases, heart attacks, and death are higher in winter than in other seasons.

Cultural Cycles

Regardless of latitudinal zone, the ideas of rest and activity are common to virtually every culture. The Christian Bible, the Bhagavad Gita, the Tao te ching and other major philosophical/religious treatises have stressed the active/passive cycle and have related it principally to the passage of time. Festivals and religious events are often held to mark significant points in a particular cycle.

Recognition of the cycle of activity/passivity is an important factor in pacing our lives. Psychologists and physiologists have found that the harder you work, without a recuperative break, the less efficient you become. In other words, it is possible for one person to work intensively for eight hours and actually accomplish less than one who works intensively for an hour, takes a break, works an hour, takes a break, etc. It also has been established that when studying or doing any intense mental activity, one should break about every fifteen minutes for optimal performance. Manifestations of the work/play/rest cycle are clearly part of many cultures. The five day work week with two days off serves as a standard work/rest cycle in many western societies, as do the annual vacation, the summer recess from school, and the eight hour work day. Other cultures break their days and weeks up differently. All cultures mingle work periods with formal rest or play periods, often in the form of holidays.

Some authors have speculated that societies themselves follow long term cycles. Frederic Jackson Turner believed that Americans were typical in that they went through the same recognizable stages that other nations did, from exploration through settlement, to urbanized groupings. Gibbon, another historian, postulated a similar cycle for all nations and used the rise and fall of Roman society as an example. We see vestiges of such philosophical points of view today in popular discussions of the rise and fall of Great Britain and other nations as world powers. Whether or not the idea is correct, it does suggest that the search for cultural analogs of repetitive physical and biological processes is a continual one.

Geological Cycles

Geological events are most often thought of as phenomena that involve millions of years for completion or, if repetitive, separated by long time periods. But there are shorter period events, those within the lifetime of individuals or certainly of specific societies, that have a strong impact on human activity. The term *cycleology,* coined by Elias, is used to refer to the study of geologic cyclic phenomena.

Certainly, a dramatic event and one that occurs in broad, but unfortunately not predictable, cycles is volcanic activity. It is an *aggradational* process, like folding and faulting, that creates *relief* on the earth's surface. A recent and familiar example was the Mt. St. Helens eruption in southern

Washington. The 1980 eruption was another in an apparent series that has about a 100 year frequency. The fact that people did not believe that it could happen, despite geophysical evidence, or that somehow they could avoid the ensuing disaster, is attributable to short memories and abiding faith that unless something happens frequently it is unlikely to happen at all.

Of less spectacular, but more significant importance, are other geological processes that produce topographic change. We mentioned aggradive processes above. In opposition is *degradation,* the destruction of relief. Some elements of these processes occur over short enough time spans and in small enough areas that we can actually comprehend their impact and plan to either aid or retard their actions. We can even use the concepts for our own benefit. For example, contour plowing is designed to retard erosion (degradation). The creation of terraces and landfills are designed to promote soil and earth build up (aggradation) to create more level land.

Longer term geologic cycles also affect human behavior and activity. Significant shifts in sea level have occurred during recorded time. Large parts of the Mediterranean coast, today under water, once were dry land. The land supported cities, harbors and life; today the area is marked only by underwater ruins of interest to archaeologists. Other sea level changes were associated with land bridges between Siberia and Alaska; bridges had a pronounced effect on the migratory patterns of early humans as well as coping strategies developed to deal with northern latitude environments.

While the long term geologic cycles associated with continental drift are well beyond human history, and perhaps beyond our ability to comprehend as a future, the affected human history will still influence contemporary life. The areas where continental plates grind against each other as they move are also the zone where we find frequent earthquakes. As we learn more about plate movement, we also will learn more about the causes of earthquakes. Such information should be useful in avoiding or at least preparing for such events so that they cause less catastrophic damage and loss of life. It is unlikely that absolute dates and times for earthquakes will be predictable, only general periods of danger. As with Mt. St. Helen's, people will probably not heed such general warnings, nor can all life and economic activity cease on the probable, but uncertain, chance that disaster will occur. These cycles of events are just too difficult to integrate into human experience—at least so far.

Concluding Thoughts

Clearly a knowledge of recurrent events, such as tides and seasons, is important in the more efficient gathering and production of food. Those who understand natural cycles have a competitive edge over those who do not. Historically, individuals who could read and interpret these "signs" were considered to possess magical powers. Often they became witch doctors and priests. As their secrets became public knowledge and everyone began to understand the importance and workings of natural cycles, these special classes lost their role, but they left behind both their methods and observations.

Today, we are beginning to understand better how our bodies are affected by various external stimuli in developing or maintaining biological cycles. A number of practical applications have appeared. For example, knowledge of circadian rhythms has allowed medicinal chemists to prescribe the proper time of day to take drugs in order to improve their effectiveness and limit adverse side effects. Psychologists have made professions of interpreting the impact of light, sound and temperatures on behavior, especially work, rest and employee productivity, and have suggested ways in which such observations can be put to use.

What is clear is that knowledge of cycles, and the relationship between external and internal events allows rudimentary prediction of behavior. This in turn leads to an ability to influence behavior by manipulating either the environmental stimuli, or by modifying behavior to match the environment.

Knowledge and awareness of the role of repetitive physical and biological events is an element in our ability to understand the man/land relationship. However, such a knowledge really focuses only on a part of the equation. To begin with the "man" portion, we need an understanding of the physiological nature and limits of the human body itself.

Terms

Aggradation. Processes that lead to the creation of relief on the earth's surface.

Annual. Plants that complete their life cycles in one growing season.

Aschoff's Rule. For creatures active during the day, increased periods of light will increase their activity while prolonged light will decrease the activity of nocturnal creatures.

Biennial. Plants that require growing seasons to complete a life cycle.

Biorhythms. Naturally occurring cycles of intellectual, emotional and physical behavior whose repetition can provide a basis for predicting optimal times for various human activities.

Biological Cycles. Naturally occurring repetitions of active and passive stages in the entire range of biological activity. The most obvious are sleep and wakefulness, activity and rest, deep sleep and rapid eye movement (rem) or dreaming.

Biosphere. The layer of earth and atmosphere in which life predominantly occurs. It includes only a shallow layer above and below the earth-atmosphere interface.

Chronobiology. The study of the timing or periodicy of various biological cycles.

Circadian Cycles. From the Latin *circa* (about) and *dia* (the day). Twenty-four hour cycles that occur regardless of light or darkness.

Crop Calendars. Displays of time periods during which the life cycle of agricultural plants is completed. These devices are used for scheduling of activities to correspond to phenological events.

Cycleology. The study of cyclic phenomena, like sedimentation, erosion, and vulcanism, in geology.

Cycles. A complete set of events that always repeat in the same sequence over about the same time period.

Degradation. The destruction of relief by weathering, erosive, and mass-wasting processes.

Diastrophism. Movement of the earth's crust. Includes such things as formation of mountain ranges and movement of plates.

Diurnal. Daytime cycles. Activity occurs primarily during the day rather than at night (nocturnal).

Endogenous. From the Greek *endo-* (within) and *genesis* (origins). Activity, cycles or biological behavior that is initiated by elements within the body.

Equinox. This is the time of year when the sun is directly overhead at the equator and day and night are equal length. Occurs twice a year and marks the transition to Spring (a period of increasing light time) and Fall (a period of decreasing light time).

Exogeneous. From the Greek *exo-* (external or outside) and *genesis* (origins). Behavior or activity that is initiated by external stimuli.

Frequency. How often an event occurs. The frequency is stated in occurrence/time units such as once/year.

Greenhouse Effect. Trapping of long wave (earth emitted) radiation by water vapor and carbon dioxide. Temporary retention keeps earth-atmosphere temperatures much higher than they would be without an atmosphere partially composed of these two gases.

Growing Season. Normally considered the time between the last and first frost in an area. The length of growing seasons is different in different climatic regions; tropical realms always have conditions above freezing while polar regions have below freezing conditions every night the year around. Some plants can tolerate freezing conditions for a short period of time after emergence and during the harvest season.

Nocturnal. Nighttime processes or cycles. Activity occurs primarily at night rather than during the day (diurnal).

Perennial. Plants that have a life cycle longer than two growing seasons. Most grow, flower, fruit and then go into dormancy after each season.

Periodicity. Recurrence of an event at regular intervals. The interval is measured in an appropriate unit of time.

Phenological Events. The life stages of plants. These stages occur at approximately the same time each year for a given species unless weather conditions prove prohibitive.

Relief. The difference between the highest and lowest point in an area.

Revolution. Movement of the earth around the sun. The time required for one revolution is 365 days, 5 hours, 48 minutes, 45.51 seconds.

Rotation. Movement of the earth around its axis which is inclined $23\frac{1}{2}°$ to the vertical. The amount of time required is 23 hours, 56 minutes and 4.09 seconds.

Seasons. Divisions of the year based on earth-sun geometry. Solstices (sun overhead at the Tropic of Cancer or Tropic of Capricorn) mark the boundary between Spring and Summer or Fall and Winter. The Equinoxes (sun overhead at the Equator) mark the boundary between Winter and Spring or Summer and Fall.

Solstice. From the Greek *sol* (sun) and *stitium* (to stand still). It is the time of year when the sun is overhead either at the Tropic of Cancer (summer) or the Tropic of Capricorn (winter).

Sunspot Cycles. Cooler areas on the surface of the sun that occur in pairs along the sun's equator. They increase or decrease in numbers of a 22.2 year cycle and have been suggested as an influence on the earth's weather and climate. They definitely affect radio signal transmission. There is also a 180 year cycle that is referred to as the Jupiter effect.

Synodic Time. This relates to the lunar (moon) period of 29.5 days from one stage to its recurrence.

Zeitgeber. A German word meaning "time setter." Used with reference to "setting" our biological clocks with cues such as light, heat, or endogenous chemicals.

References

Adderley, E. E. and E. G. Bowen. "Lunar Component in Precipitation Data." *Science* 137: 749, 1962.

Aschoff, J. "Circadian Rhythms in Man." *Science* 148: 1427, 1965.

Bradley, D., M. Woodbury and G. Brier. "Lunar Synodical Period and Widespread Precipitation." *Science* 137: 748, 1962.

Brezowsky, H. "Morbidity and Weather" in S. Licht (ed.) *Medical Climatology*, Phys. Med. Library. Baltimore: Waverly Press, 8: 358, 1964.

Campbell, D. and J. L. Beets. "Lunacy and the Moon." *Phychological Bulletin* 85 (5):1123, 1978.

Cloudsley-Thompson, J. *Rhythmic Activity in Animal Physiology and Behavior*. New York: Academic Press, 1961.

Colquhoun, W. *Biological Rhythms and Human Performance*. New York: Academic Press, 1971.

Halberg, F. "The 24 Hour Scale; A Time Dimension of Adaptive Functional Organization." *Perspectives in Biology and Medicine* 3: 491, 1960.

Hilts, P. "The Clock Within: The Body's Internal Cycles Affect Everything from Blood Pressure to Decision Making." *Science* 80, December, 61, 1980.

Klein, R. and R. Armitage. "Rhythms in Human Performance: 1½ Hour Oscillations in Cognitive Style." *Science* 204: 1326, 1979.

Luce, G. G. *Body Time*. New York: Bantam Press, 1972.

Mills, J. H. ed. *Biological Aspects of Circadian Rhythms*. New York: Plenum Press, 1973.

Mironovitch, V. "Planetary Position Effect on Shortwave Signal Quality." *Electrical Engineering* 71: 421, 1952.

Palmer, J. and F. A. Brown, Jr. *An Introduction to Biological Rhythms*. New York: Academic Press, 1976.

Raloff, J. "Biological Clocks: How they Affect your Health." *Science Digest*, November 62, 1975.

Sandow, S. *Durations*. New York: Quadrangle/New York Times Book Co., 1977.

Saunders, D. *An Introduction to Biological Rhythms*. New York: John Wiley and Sons, Inc., 1977.

Still, H. *Of Time, Tides and Inner Clocks*, New York: Stackpole Books, 1972.

Environment and *Homo Sapiens*

All species exist within what is called a *range of tolerance* for environmental factors. This range is based upon the biological limits beyond which the species cannot continue to survive or reproduce itself without assistance. The major determinates of the range are temperature, water, and sensitivity to light. There are biochemical conditons as well, but we can ignore these for the moment.

The range of tolerance consists of a continuum from maximum to minimum, with *optimum* conditions marked as those under which the organism can thrive and multiply. The optimum conditions for any species can be determined by the environments in which it is found in the richest variety and numbers. At minimum or maximum limits, an organism occupies only favorable locations, is often stunted, and is found in only limited numbers. It is at the margin of its range of survival. Any greater stress may lead to exclusion of the species from that environment. We often see this in areas where grasslands and forests merge. Each grass or tree species is at the margin of its range of tolerance (Fig. 3.1). These marginal zones are called *ecotones*.

Physiological stresses

Humans also live within a range of tolerance determined by physiological characteristics. External temperature and moisture values, along with oxygen supply and the chemical balance of the blood, determine what the range will look like at a given moment. Along with other warmblooded mammals, humans have special mechanisms which allow them to adjust to a much wider environmental range than many other species. They adjust to internal and external factors in order to reach *homeostasis* or physiological well-being within the constraints imposed upon them. The ability of the body to independently balance its needs for temperature, water and other functions of life by self-regulating mechanisms provides the mechanism for attaining homeostasis. Thus, if the body becomes too cold, heat is generated, perhaps by shivering. If it is too hot, the rate of perspiration is increased to allow evaporative cooling of the skin and blood.

Figure 3.1. An *Ecotone* between grasses typical of the Great Plains and coniferous trees typical of the lower elevations of the Rocky Mountains. Trees cannot survive in the grasslands and plains grasses cannot compete favorably with pines at higher elevations.

Water and Temperature Stresses

Human environmental limits in an unclothed state in the absence of any attempts at adjustment are determined by the makeup of the body and its autonomically controlled mechanisms of adjustment. We are sixty-six percent by weight water or fluid. Variation from this percentage of one percent or more will cause discomfort. Excess fluid will be eliminated by increased perspiration or increased urination. The body will begin to automatically adjust to a water deficiency by drawing water from stored fat, tissues, or the blood. Self-inflicted dehydration results.

A loss of ten percent of bodily fluid will so change chemical blood composition that there is insufficient flow and energy left to provide muscle action and a person cannot walk. If losses reach twenty percent, death is imminent unless the fluid is quickly replaced. Fluid balance, then, occurs over a rather narrow range and is an important aspect of physiological well being.

We can live many days without food, but only short periods without water. Water is crucial to internal homeostasis; it lubricates the organs and cells, maintains blood flow, and carries nutrients and oxygen to the brain and organs.

The human body also is very sensitive to small fluctuations in temperature. Normal or average deep body (core) temperature is 98.6°F (37°C). Slight percentage variations from equilibrium temperature cause immediate discomfort. A temperature of only 100°F (38°C) is considered cause for concern in adults and must soon be reduced if the body is not literally to burn itself up. Likewise, a rapid decrease in core temperature may lead to hypothermia and death.

Temperature stresses are least in a small thermally neutral zone extending from air temperatures of 25°C (77°F) to 27°C (80°F). A nude person at rest is in thermal equilibrium with the environment between these temperatures and all adjustments to temperature fluctuations are physiological (muscular, cardiovascular and metabolic adjustments). Outside of this zone, changes are behavioral and include such adjustments as changing the covering on the outer surface of the body, postural changes, food ingestion, water consumption, and diurnal/nocturnal activity patterns.

Biochemical Constraints

Humans are equally vulnerable to a wide range of biochemical variations associated with diet and the environment. The body's chemical balance is very complex. We are just beginning to understand how critical to normal activity and behavior many of its chemical constitutents are.

There are a wide range of nutrients associated with foods that are essential to normal functioning of homeostatic mechanisms. Some essential nutrients, their physiological role, and source foods are shown on Table 3.1. For a long period of human history, their provision was linked to foods gathered or grown around the immediate home and thus to foods adapted to surrounding environmental stresses. In the tropics this meant a largely vegetarian diet; in colder latitudes more meat was necessary, while in coastal areas, fish and shellfish were dominant dietary staples. It is apparent native peoples developed internal biota (bacteria in the intestines) to process foods in such a way that necessary vitamins and minerals would be extracted that might otherwise be lacking in the diet.

Recent research has shown that many chemicals are important for normal bodily functioning and behavior. Of these, the most talked about are vitamins, but there also are chemical substances such as the peptides found primarily in brain functioning, and electrolytes found primarily in blood, that

Table 3.1. Nutrients

Vitamin	Effect	Source(s)
A	Presumed to be an infection fighter since its level drops during infection. Important to health of epithelial (body lining) tissue, especially nose and throat. Enhances immunity to various viruses. Aids vision.	Carrots, broccoli, cabbage, tomatoes, liver, eggs, milk
B group		
B_1 (thiamin)	Protects the nervous system. Large injections have been used to overcome the damage of alcoholism.	All are found in abundance in green vegetables especially spinach, and asparagus but also in bananas, carrots and onions.
B_2 (riboflavin)	Aids in healing skin sores, itchy dry skin and problems associated with the tongue.	
B_3 (niacin)	Converts food to energy, aids the nervous system, prevents loss of appetite, pelagra and dermatitis. It also has been related to anxiety, depression, fear, confusion, anemia, forgetfulness, and paranoia.	
B_5 (pantothenic acid)	Needed for the body's ability to use carbohydrates, fats, and proteins.	
B_6 (pyridoxine)	Can be substituted for cortisone for skin problems when applied locally as a salve. Useful in overcoming nervousness, depression, asutistic behavior. Acts as a co-enxyme.	
B_{12} (cyanocabalamin)	A lack causes pernicious anemia.	
C	Has been found effective against hepatitis, measles, mumps, herpes, meningitis, the flu and various bacterial and fungal infections. Should be used with bioflavinoids and rutin.	All citrus fruits
D	Can be synthesized in the body when the skin is exposed to sunlight. Promotes the absorption of calcium and maintains the balance of calcium and phosphorous. Prevents defective bone growth such as rickets.	Various fish oils
E	An anti-oxidant. It has been used to offset menopause problems, shingles, angina, hemorrhoids, rashes, burns and leg cramps as well as phlebitis.	Most green vegetables
F	Seems to be related to the nervous system and brain activity.	Seafood, cold water fish, olive, peanut, corn, sesame, and soybean oils.

play equal if not more critical roles in the proper functioning of the body. Many of these seem directly related both to diet and changes in the body brought about by environmental stress.

Vitamins are seemingly well understood (Table 3.1). The United States government has established what is termed the MDR (minimum daily requirement) for most vitamins. However, this is a minimum, the very least necessary to prevent serious illness or disease, and these limits normally are met in any reasonable diet. The question of what is the optimum amount of vitamins necessary for good health has yet to be agreed upon. The result is that there is much controversy and many fads stressing one set of vitamins or another as a cure-all.

A more recent discovery and much less well understood element in human health and behavior are hormones produced by the brain. The better known of these are listed in Table 3.2. These are internally generated (endogenous) chemicals that affect or control mood, pain, and even the regulation of body temperature.

Table 3.2. Selected Brain-produced Hormones

	Endorphins and *Peptides*
Substance P	Involved in transmitting pain signals to the brain.
Factor S	Apparently related to sleep promotion as its presence increases sleeping time.
DOET	Acts like a double martini.
LRH	Increases the sperm count.
Beta Blocker	Surpresses the flow of adrenalin, thus blocking the receptors for anxiety reactions. Beneficial for persons with heat ailments.
Beta Endorphin	Aids in the suppression of pain. Similar to morphine.
Nootropyl	Increases intelligence and information flow between the right and left hemispheres of the brain.
Norepinephrine	Increases aggression.
Sertonin	Increases passivity.
ACTH	Manufactured by the pituitary gland. Found in all nerve cells in the brain. An aid to learning and memory. Similar to vasopressin.
Bombesin	Can turn warm blooded animals into cold blooded animals when injected into the brain. Turns on the sympathetic nervous system (stress responses).
Bradykinin	Very small amounts cause intense pain when injected. May be the most painful substance known.
Cholecystokinin	Concentrated in the cortex. Seems to be related to the signal to eat.

Finally, there is the matter of electrolytic balance in the blood—the blood's pH or balance between acidity and alkalinity. The chemicals most directly involved are potassium and sodium, but other elements such as calcium and manganese also are involved. Rather minute amounts (ions) of each are involved, yet their impact on human behavior and health is immense, even if their mechanisms are not yet well understood.

For example, under an imbalance of sodium and potassium, we may note a kidney malfunction, or the effects of extreme heat and high fluid losses. Such losses may not only indicate physiological failure, but they also seem to be associated with causing such failures in otherwise healthy persons. In addition, there may be severe mood or behavior changes. A loss of potassium ions may cause weakness and listlessness with an impairment of muscle movement and tissue deterioration. While the full implications of electrolytic balance in the blood are still being studied, it is clear that these are tied closely to general homeostasis and thus essential to successful survival.

In the context of the physical environment, it is important to note that all of these chemicals and vitamins can be affected by thermal stress, and dietary patterns. For example, among the Eskimos of Greenland, there is a disease known as arctic hysteria. It is due to the prolonged winter darkness associated with the region, resulting vitamin deficiencies (primarily B and D) and electroylte imbalances. Behavior is bizarre, but not dangerous. To cope with this common occurrence, the local culture has integrated this period of the year and its related behavior into a kind of religious context. Those affected become respected and treated with care and kindness, not approbation and fear.

The question arises then as to how many cultural behaviors are related to environmental affects on diet and resulting behavior changes? For example, does fasting, or its converse Thanksgiving, relate to seasonal availability of food? Are dietary prohibitions related to seasonal foods? Are food prohibitions related to health and behavior consequences?

Diet

Biochemical constraints require than an adequate diet is possible under prevailing environmental conditions, or, if an environment is not capable of sustaining human life, energy in the form of food must be imported. Specific dietary needs depend upon many factors, including age, health status, environmental stress, and activity of the individual. We can get a general idea of typical recommended daily bulk requirements for the average 23–50 year old male living in the United States from the first three rows on Table 3.3. Younger males, pregnant women, very active people, those under stress, and those with certain health problems would probably require a higher caloric intake. The remaining last two columns on the table show materials used to process energy sources.

Table 3.3. Metabolic Requirements for Life Support of One Man Per Day Required Supplies

	Energy (Cal)	Food (g)	Oxygen (ml)	(g)	Water (g)	
Carbohydrate	1200	300	244	342	1000 (food)**	
Protein	320	80	90	114		
Fat	1350	150	300	420	1500 (liquid)	
Total	2870*	530*	634	876	2500	3.906 kg

*These calculations assume complete digestion, absorption, and oxidation of this food material. Such complete combustion is not achieved with natural foods. Hence they will weigh somewhat more than here indicated.

**Natural food is about ⅔ water. Thus, to obtain 500 g of protein, carbohydrate, and fat, 1500 g of steaks, peas, and bread, which contains 1000 g of water, must be consumed.

Ursual T. Slager, *Space Medicine*, Copyright © 1962, p. 266. Adapted by permission of Prentice-Hall, Inc., Englewood Cliffs, NJ.

Dietary patterns for other countries differ, often markedly, from these figures. For example, Norwegian (cold climate) males need an average caloric intake of 3400 calories/day; Filipinos (warm climate) need only about 2400 calories/day. Some Guatemalan villagers subsist on 1500 calories/day, as was the case in pre-war Japan, without apparently deleterious effects on mental and physical well-being.

The time of food intake and the amount per feeding also vary widely. For example, three nutritionally complete meals per day are the American ideal. Other societies favor one meal over others as the most complete and main meal of the day; Dubos points out that the mid-day meal serves this function in France. Some groups eat almost continuously during the day while foraging; others eat only once. Anthropologists suggest that time/amount eating habits may be adaptive, or simply the result of conditioning of children within societal groups; learned patterns are often retained throughout life, even if individuals move to an entirely different environment.

Sources of Dietary Elements. Energy and other needed elements are derived from carbohydrates, fat, and protein. Approximate percentages of each for adult Americans are shown on Table 3.3.

Foods with a high starch and sugar content provide carbohydrates with an energy equivalent of about 4 kcal per gram. Root crops, like potatoes, yams, and manioc, have a high starch content (upwards of twenty percent), as do flour (40–50 percent) and cereals (70–80 percent). High sugar content foods include, again, root crops, certain dried fruits, honey, dates, and fruits. While about fifty percent of the total daily energy supply is derived from carbohydrates in western societies, the percentage rises to around seventy percent for most African, Asian and South American countries.

Fats have twice the energy content/gram of either carbohydrates or proteins. They provide the greatest supply of reserve energy in animals. Fat is stored under the skin, particularly among peoples who live in cold environments serving as an insulating blanket, and around the viscera. Fats are derived from oils in foods like nuts, meat, and milk. Fats comprise about thirty-five percent of the daily energy intake among western societies. Among Eskimo populations, the percentage may reach eighty.

Proteins comprise about half the dry weight of humans; about one-third of the total is in muscles, twenty percent in bones and cartilage and the rest distributed throughout the tissues. Proteins provide amino acids which allow enzyme function, hormone synthesis (Table 3.2), antibody formation, tissue construction, regulation of certain bodily processes, and they are a source of energy. Animal foods, such as red meat, poultry, milk, cheese, eggs and fish, are suppliers, as are certain grains like soybeans. Most grains are low in protein (9–14 percent), however, but are often consumed in large amounts and thereby serve as a significant protein source in some countries. Western societies derive about twelve percent of the total daily caloric intake from protein.

Historical Changes. It is clear that we have spent well over ninety nine percent of our existence as a species in a seemingly primitive food provision mode (Table 3.4). Energy was obtained principally from a) foodstuffs limited to what the *biome* was capable of providing and b) the lowest levels in a foodchain. Dietary requirements, the skills needed to obtain proper nutrition, and the actual diet used to supply daily energy requirements changed little over the early millenia of human existence; major changes which have occurred took place during the last several hundred years.

Clements emphasizes this point in this three part division of basic diet patterns. He suggests that primitive man practiced hunting and gathering exclusively and existed on principally a flesh food diet. By about 10,000 years ago, grains began to be added to the list of food staples and formed as much as ten percent of the daily intake.

The beginnings of the historical period in the Middle East and Western Europe were marked by a shift to a mainly vegetable diet. Up to ninety percent of total diet weight was made up of cereals. Domestic agriculture of this type allowed maximal energy production from a small area. This pattern still exists in parts of Asia and Africa today; intensive rice production serves as an example.

During the last 1000 years, but particularly the last 350 years in industrialized societies, dietary patterns have shifted to a mixed vegetable/ animal form with vegetables forming some 30–70 percent of the diet. Again, the influence of efficient energy usage for larger populations is important. One acre (0.5 ha) will produce 4,000,000 kcal of potatoes but only 350,000 kcal of beef.

Table 3.4. Generalized Time Scale for Selected Events That Improved Carrying Capacity*

Years Before The Present (BP)	Generations (25 yrs/gen)	Agricultural Event	Various Groupings of People
10^6	5×10^4	Hunter/Gatherer	Widely separated bands of less than 90
10^4	5×10^2	Beginnings of domestic agricultural communities	Semi-permanent nomadic bands of less than 250
5×10^3	2×10^2	Advent of irrigated agriculture	Some citites of 1×10^5, principally small towns of less than 300
1×10^2	4	Beginnings of scientific search for better crop varieties	Some cities of 5×10^5, numerous cities of 1×10^3
50	2	Widespread Introduction of mechanical farming in western societies	Some cities of 1×10^6, many cities of 2×10^5, many towns of 1×10^3
20	<1	Widespread use or chemical fertilizers and pesticides along with mechanical agriculture	Some cities of 8×10^6, many cities of 1×10^6, decreasing number of towns of 1×10^3
0	<1	International trade obligatory to maintain adequate nutrition even in some western countries	Some cities of 16×10^6, many cities of 5×10^6, great number of towns of 1×10^5, few towns of 1×10^3

*We are aware of current arguments by Boserup and others on the evolution of agricultural practices. This table is devoted solely to some developments that have had an effect on carrying capacity and the efficiency of biome energy use.

Homeostatic Mechanisms

Continuation of life requires that bodily homeostatic mechanisms maintain and monitor the balance of all the above factors as well as many additional functions. The natural environment adds to the problem by altering variables that require reaction by the body. For example, extreme cold or heat demand changing metabolism and thus varying food consumption. High altitudes reduce available oxygen that affects oxidation of food and ability of muscles to function. Regardless of the environment, the body must maintain a relatively constant fluid and temperature state in its core.

The body attempts to maintain its equilibrium in any environment at the lowest possible energy cost to itself. In order to do so, it must read the external environment properly and then adjust to any physical stress as quickly and efficiently as possible.

The body may gain or lose heat by either biological or environmental means. The important thing is that its core temperature is maintained at an average of 98.6°F (37°C). Maintenance of this temperature requires a balancing act that is performed primarily by the body's homeostatic mechanisms, but in severe environments it must be aided by clothing, housing and diet. We will consider humans and the physical environment in this chapter and then proceed to the latter three items in subsequent chapters.

Thermal Equilibrium

The energy exchanges that occur between the body and the environment can be expressed in symbolic form as follows:

$$\pm B = M \pm (\pm R \pm C \pm E)$$

where:

- B = Homeostatic balance
- M = Production rate of total metabolic heat
- R = Net radiational heat exchange
- C = Conductive/convective/advective energy exchange
- E = Rate of heat loss or gain by evaporation or condensation

The result of physiological mechanisms that act to maintain thermal balance are represented by the *Basal Metabolic Rate* (BMR). Simply stated, this is the rate at which the body generates heat at rest; heat produced is just enough to maintain the processes of life exclusive of any stress due to heat or cold, exercise, or allementation. In fact, the simple act of digesting food raises the BMR an average of ten percent or more.

Table 3.5. Approximate Number of Calories Produced (M) for Different Activities by an Adult of $1.7M^2$

Metabolic Rate KCAL/HR	Activity
680	Raquetball
590	Digging a hole to plant a tree
160-357	Household chores
90-130	Office work
340	Walking at average pace
500	Walking fast with load
120	Standing
70	Sleeping

Table 3.5 illustrates heat produced in *calories* for various normal activities. Bear in mind that body size, age, sex and the temperature under which these activities occur can change caloric rates drastically. For example, it is estimated that it takes a minimum of 5,000 calories per day just to maintain life under arctic conditions. In contrast, less than 2,000 calories per day are required to maintain BMR in the hot and humid tropics.

The many ways in which the body can gain or lose heat to its surroundings are shown in Figure 3.2. We start by noting that heat always moves from a warmer to a cooler medium. Thus, any object hotter than human skin temperature (about 91°F or 33°C) will transfer heat to the body. In fact, thermal stress begins well before air temperature reaches 91°F since at 91°F the body loses its ability to dissipate its internally generated heat. While air temperature is certainly the most obvious element in heating or cooling the body's surface as Figure 3.2 illustrates, there are many other ways in which we may be heated or cooled.

When air is moving, either as natural currents or because a fan causes it to move, it produces increased cooling; moisture on the skin is evaporated and in the process carries away body heat. The technical term is *latent heat of evaporation*. When air is very humid, this cooling process is less efficient and we feel much warmer than we might at equal air temperatures and less humidity. Conversely, in very dry air the rate of evaporation, and, thus, the cooling process, is accelerated; we can tolerate a much higher temperature in a dry desert. Bear in mind, however, that the body is still under abnormal thermal stress and that prolonged exposure to either condition can lead to difficulties. When the body comes into contact with solid objects such as building walls, chairs, benches, floors, etc., it will either lose or gain heat by conduction at a rate that depends upon the difference in temperature between the skin and the object. Typical partitioning of heat loses under selected conditions is shown on Table 3.6.

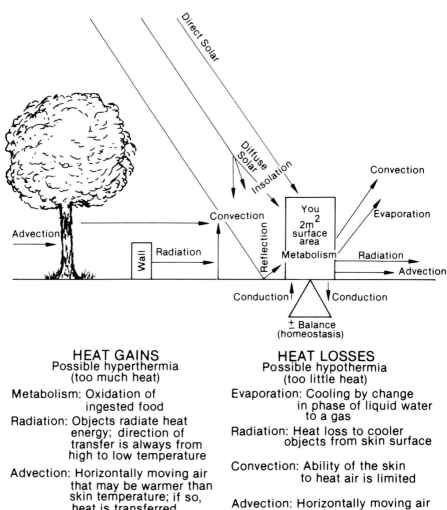

Figure 3.2. Energy exchanges between humans and the environment.

Table 3.6. An Example of Heat Loss to the Environment

Room Temperature	Radiation	Percentage Loss by: Convection	Evaporation
Comfortable (25°C)	67	10	23
Warm (30°C)	41	33	26
Hot (35°C)	4	6	90

Reprinted from E. Folk. 1974. Textbook of Environmental Physiology (2nd Ed.). Philadelphia: Lea and Febiger.

Table 3.7. Summary of Human Responses to Thermal Stress

To Cold	To Heat
Thermoregulatory responses	
Construction of skin blood vessels	Dilation of skin blood vessels
Concentration of blood	Dilution of blood
Flexion to reduced exposed body surface	Extension to increase exposed body surface
Increased muscle tone	Decreased muscle tone
Shivering	Sweating
Inclination to increased activity	Inclination to reduced activity
Consequential disturbances	
Increased urine volume	Decreased urine volume. Thirst and dehydration
Danger of inadequate blood supply to skin of fingers, toes, and exposed parts leading to frostbite	Difficulty in maintaining blood supply to brain leading to dizziness, nausea, and heat exhaustion
	Difficulty in maintaining balance, leading to heat cramps
Increased hunger	Decreased appetite.
Failure of Regulation	
Falling body temperature (Hypothermia)	Rising body temperature (Hyperthermia)
Drowsiness	Heat regulating center impaired
Cessation of heartbeat and respiration	Failure of nervous regulation terminating in cessation of breathing

Adaptation of a table from *Review of Research, Arid Zone Research* X, (c) UNESCO 1958.

When either heating or cooling of the body occurs faster or longer than bodily mechanisms can adjust to, a serious threat to life occurs (Table 3.7). These conditions are called *hypothermia* and *hyperthermia*. Each can occur in what otherwise would appear to be normal conditions and each have caused needless deaths simply because people were unaware what was happening and failed to take remedial measures. Familiarity with the nature of hypothermia, hyperthermia and their symptoms further illustrate how the homeostatic mechanisms of the body work.

Hypothermia

Hypothermia, once called "exposure," is the rapid loss of body heat, loss so rapid that the body cannot generate sufficient heat internally to compensate for losses. It is subtle, almost painless, and almost always fatal if not treated immediately.

Hypothermia can occur in sunny, warm weather. It can happen when you are hiking or riding a bicycle and a light rain soaks you and your clothing. The resulting chill can lead to hypothermia—and often does among hikers and bikers in the mountains—even in the summer. Hypothermia begins by the loss of body heat to the environment. This happens when you sit on a cold rock and the body gives up its heat to try and heat the rock to body temperature. It occurs when the skin is wet and the moisture evaporates taking away needed body heat. Perhaps the worst condition, since it maintains an almost constant drain on body heat, is wet clothing in a slight breeze.

The symptoms of hypothermia include the following:

1. Core temperature, 99°F to 96°F (37°C to 36°C): shivering becomes intense, ability to do simple tasks is slowed.
2. 95°F to 91°F (35°C to 33°C): skin tone pales, shivering becomes violent, speech is impared.
3. 90°F to 86°F (32°C to 30°C): muscular rigidity replaces shivering, thinking is dulled.
4. 85°F to 81°F (29°C to 27°C): become irrational and may drift into a stupor, pulse is slow.
5. 80°F to 78°F (26°C to 25°C): unconsciousness occurs, reflexes cease to function.
6. Below 78°F (26°C): condition may be irreversible and death is likely.

Physiologically what is happening is that blood is withdrawn from the extremities (fingers, toes, nose, and even the peripheral areas of the brain, the neo-cortex). This blood is pooled in the *thoracic cavity* to maintain a constant body temperature. Blood is not sent to the skin and periphery of the body, as there it would lose heat to the environment. As the blood is pulled away from the periphery and the brain, speech and thinking are impaired, and the individual no longer really cares what is happening. It would be nice just to sleep (a kind of permanent hibernation).

Wind chill is caused by moving air speeding up the rate of evaporation and thus of heat loss from the exposed surface areas of the body. At moderate temperatures, a mild wind is very cooling and even beneficial. As the air temperature drops and the body's metabolism is increased just to maintain life, any increase in the rate of heat loss puts much greater stress on the body. Thus even a mild wind at temperatures of 40°F (14°C) can quickly overcome the body's ability to maintain life if no food or shelter is provided. Wind chill can rapidly accelerate hypothermia even when the conditions otherwise appear

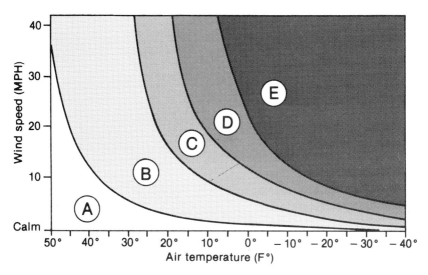

Conditions in each zone

A Comfortable under normal conditions.

B Very cold. Travel becomes uncomfortable on overcast days.

C Bitterly cold. Travel becomes uncomfortable even on clear, sunny days.

D Exposed flesh may freeze, depending upon degree of activity, amount of solar radiation, and character of skin and circulation. Travel and life in temporary shelter becomes disagreeable.

E Survival efforts are required. Exposed flesh will freeze in less than one minute.

Figure 3.3. Wind chill

nonthreatening. Figure 3.3 illustrates the effect of wind chill and provides some idea of the conditions under which life could be threatened if no protection is provided.

If you are in the sun (solar radiation heat gain) and no wind is blowing, the temperature on your skin is much higher than actual air temperature. Thus, in the sun, at 50°F (10°C) your skin would react to an effective temperature of 74°F (23°C); even a mild breeze of 5 mph (2.5 mps) reduces this to 48°F (9°C).

Hyperthermia

Hyperthermia is the exact opposite of hypothermia. It is the rapid buildup of heat in the body beyond what the body is capable of dissipating naturally. It is also associated with heat prostration and heat stroke. It is similar to a prolonged, but very high, fever. The body in effect burns itself up and, in the process, places such a strain on the heart, trying to pump sufficient blood to the surface of the skin to help dissipate heat, that even a young man can die of a heart attack or heart arythmia.

The symptoms of hyperthermia include cramps, exhaustion and heat stroke. Much more symptomatic is the process of dehydration. Dehydration leads to gradual thickening of the blood, a loss of the lubrication of cells, and the necessary bathing of all tissues in fluids that keep them active and alive. In addition, the loss of fluid rapidly upsets the chemical balance of the blood itself causing toxins to form that actually poison the body through a sudden increase in concentration.

Water addition offsets the problems associated with fluid loss and also offsets the problems of heat buildup. It is important to maintain a situation in which the body and its tissues are as saturated with fluid as possible.

Water needs under various conditions are shown on Figure 3.4. These values do not take into account changing relative humidities: the higher the humidity, the less successful evaporative cooling will be and the less water

	Maximum daily temperature (°F) in shade	Available water per man, U.S. quarts					
		0	1 qt	2 qts	4 qts	10 qts	20 qts
No walking at all		Days of expected survival					
	120°	2	2	2	2.5	3	4.5
	110°	3	3	3.5	4	5	7
	100°	5	5.5	6	7	9.5	13.5
	90°	7	8	9	10.5	15	23
	80°	9	10	11	13	19	29
	70°	10	11	12	14	20.5	32
	60°	10	11	12	14	21	32
	50°	10	11	12	14.5	21	32

	Maximum daily temperature (°F) in shade	Available water per man, U.S. quarts					
		0	1 qt	2 qts	4 qts	10 qts	20 qts
Walking at night until exhausted and resting thereafter		Days of expected survival					
	120°	1	2	2	2.5	3	
	110°	2	2	2.5	3	3.5	
	100°	3	3.5	3.5	4.5	5.5	
	90°	5	5.5	5.5	6.5	8	
	80°	7	7.5	8	9.5	11.5	
	70°	7.5	8	9	10.5	13.5	
	60°	8	8.5	9	11	14	
	50°	8	8.5	9	11	14	

© Reprinted from "Physiology of Man in the Desert", by E.F. Adolph and Associates. Interscience Publishers, New York, 1947

Figure 3.4. Water needs under various temperature conditions. (Adapted From E. Adolph and Associates. Physiology of Man in the Desert. New York: Interscience, 1947.)

will actually be required. The chart shows that even at low temperatures, water intake must be maintained. Indeed, dehydration of workers and even vacationers at low temperatures is more marked than often thought.

If water is not available and you are caught in a heat stress condition without water or perhaps live in a typically warm region during the summer season, time available to you will be limited. You could manage only two days in the desert southwest of the United States, even with prudent measures. Perhaps more surprising is the limited survival time available in parts of Idaho or Washington or even western Nebraska. The availability of modern transportation and water supplies has made us seemingly immune from water requirement problems unless disaster strikes.

Altitude Problems: Hypoxia

While people who live continuously at high altitudes often have adapted to the constant state of reduced oxygen and find only limited difficulty in descending to lower altitudes, the process is not equally easy in reverse. Many individuals who arrive in high altitude locations (over 10,000 feet) have difficulty in adjusting to such condition. Many become sick and must be brought down to lower altitudes immediately (Table 3.8).

Hypoxia, or acute mountain sickness (AMS), today is more fully understood than ever before but there remains much to learn. Still, we do know the major outlines of why it is so serious and life threatening. Symptoms associated with all mountain sickness (both mild and acute) include nausea, shortness of breath, headaches, sleeplessness, a feeling of being tired and a general loss of appetite. For most people, these symptoms pass in a few days and they then are fine. For others, these symptoms may merely be the early signs of more serious complications.

Table 3.8. Selected Effects and Responses to Hypoxia

Function Affected	Reaction
Ventilation	Breathing becomes more rapid
pH of Blood	Becomes alkaline due to increased ventilation
Heartbeat	Increases but maxima possible at sea level never attained thereby decreasing total blood circulation
Red Blood Cells	Increase in number due to overproduction
Light Sensitivity	Decreases
Hearing	No Change
Thyroid Activity	Decreases

The body's homeostatic mechanisms try to compensate for reduced oxygen. Reduced atmospheric pressure also triggers critical changes in internal fluid balances. For example, reduced barometric pressure allows fluids, especially in the lungs and cranium, to fill those cavities causing pulmonary *edema* (pneumonia) and cerebral edema with resulting loss of muscle coordination. There is a reduced threshold of pain, and a loss of weight and appetite. Even minor infections may become fatal. The only solution is an immediate removal to lower altitude. Here improvement is normally quick to immediate and requires no medical assistance if it occurs soon enough.

Some researchers feel there is a definite genetic factor involved in the ability of people to adapt to high altitudes. There is no question that those who were born and have lived at high elevations have larger hearts and lungs than those born at sea level, and that these characteristics are what allow successful adjustment to high altitude situations.

Some Morphological Adaptations to the Environment

In addition to the homeostatic mechanisms that maintain biological equilibrium, there also are long-term physiological adjustments the body may make. These changes aid survival as man migrates from one environment to another. Some of these adjustments are short-term, such as adjustments to higher altitude or extreme cold or heat. Others are more fundamental and involve modifications of the blood chemistry or skin color or size and shape of the body or its features.

Among warm blooded mammals, there has been a clear adaptation of body size and shape to variations in average environmental temperatures. These are expressed in *Bergmann's and Allen's Rules.*

In 1847 Bergmann stated the following proposition:

> "Within a polytypic warm-blooded species, the body size of the subspecies usually increases with the decreasing mean temperature of its habitat."

Thirty years later, the biologist Allen notes that:

> "In warm-blooded species, the relative size of exposed portions of the body decreases with the decrease of mean temperature."

What these observations mean is that members of the same species found living in cold climates eventually evolve shorter or smaller appendages (arms, legs, ears, noses, etc.) than their relatives in the warmer climates (Allen's Rule) and that in cold climates, body size will be larger, with more mass to provide the body heat needed for survival (Bergmann's Rule). In fact, at the hot end of the range of temperature tolerance, we find that appendages become extended and body size (and thus its heat preserving mass) diminishes. These adjustments facilitate the radiation loss of body heat. The classic example is a comparison of the fennec and arctic fox, both members of the same species *(Vulpes)* (Fig. 3.5).

Figure 3.5. Some animals have a smaller ratio of surface area to body mass than other animals, which means that they lose heat less rapidly than animals with a large area/mass ratio.

The advantage of smaller mass and longer arms, legs, ears, and nose is to reduce the buildup of body heat and to aid in dispersing it over a longer body surface. Man has never lived in any environment long enough for natural evolution to provide clear cut examples of Allen's and Bergmann's rules. However, we do find some interesting behavioral manifestations that clearly equate with the rules in man's short-term thermal adjustments. For example, when we are cold, we "ball up," tuck our legs and feet in, and place our hands around our chest, often under our arms. This reduces our surface area and substantially increases (temporarily) our mass. Conversely, when we are too hot, we immediately spread out. We extend our arms and spread our legs so that we expose the maximum surface area to direct contact with the cooler air in hopes that it will carry off the excessive heat in our bodies.

While hardly scientific evidence, there also is an interesting correlation between some indigenous house styles and climate. The size and shape of houses indigenous to cold climates often take small massive shapes with a minimum of surface area. The yurt of Central Asia, the igloo of the Eskimo, the tepee of the Plains Indians, and hogans of the Hopi and Navaho are examples. Houses indigenous to the hot, humid tropics are exactly the opposite. They are built of flimsy materials (palm fronds and rattan), are largely open (of little mass) and designsd to permit maximum air flow over the house and its occupants. Similarly, in the Mediterranean, where summers can be very hot and winters are mild, we often find houses with many wings and extensions as well as open windows and verandas for maximum exposure to cooling breezes. Examples of such housing, along with analytical descriptions, are found in Chapter Six.

Terms

Allen's Rule. Simplified, this rule states that in warm-blooded species, the arms, legs, nose and ears (appendages) will be small with low average temperatures and larger where the average annual temperatures are warmer or hot. The appendages thus act as "radiators" to lose body heat in warm to hot climates and to conserve body heat in the colder climates.

Basal Metabolism Rate (BMR). The amount of energy (calories) required just to maintain life when the body is at complete rest.

Bergmann's Rule. Simplified, this rule says that warm-blooded species have a more compact (rounded) and larger shape and size when the average annual temperature is cold and are more angular (leggy or beaky) when the average annual temperatures are warmer or hot.

Biome. A structural and functional model of the interaction of physical and biological elements in an area. The amount and type of foodstuffs are different in each biome. Biomes collectively make up the biosphere.

Calories. There are two common uses of this word. The small calorie is a measure of the amount of heat needed to raise the temperature of 1 gram of water 1 degree centrigrade; it is used as a physical measurement. The large calorie or kilogram calorie (kcal) is used for measuring the energy produced by food when oxidized in the body or as an amount of food able to produce one large calorie of energy. The large calorie equals 1000 small calories or one kcal.

Ecotone. A transitional community composed of species at the margins of their tolerance ranges.

Edema. Fluid build up in body tissue that may cause swelling and pressure. It is life threatening when it occurs in the brain (cerebral edema) or the lungs (pulmonary edema).

Endorphine. From the Greek *endo* (within) and morphine. It is a morphine-like hormone naturally produced by the body's pituitary gland. It is a natural defense against pain.

Homeostasis. A state of internal physiological balance or equilibrium of functions and chemical conditions designed to maintain life.

Hyperthermia. A rapid rise in body temperature beyond what the homeostatis mechanisms can cope with. Leads to heat stroke and death if not reversed.

Hypothermia. A rapid loss of body heat beyond what the body's natural system can replace or overcome. Leads to death. Also referred to as "exposure."

Hypoxia. Reaction to inadequate oxygen supply to the lungs and tissues.

Latent Heat of Evaporation. A measure of the amount of heat needed to convert liquid water to water vapor at a specific temperature. The heat is retained by the vapor until released in the process of condensation.

Optimum Conditions. The value of a factor or factors that is associated with maximum efficiency of life processes for an organism.

Peptide. Chemical substance found in the brain that is identical to hormones found throughout the body.

Range of Tolerance. The physical limits (especially heat, water and light) beyond which (minimum, maximum) any living organism can survive and reproduce itself.

Thoracic Cavity. The chest area that contains the most vital organs to life (the heart and lungs). Blood (warmth) will be pooled here to preserve life when cold threatens and the heart will beat faster to pump the heated blood away if hyperthermia is the threat.

References

Baker, P. T. "The Biological Adaptation of Man to Hot Deserts." *American Naturalist* 92: 337, 1958.

——— and J. S. Weiner, eds. *The Biology of Human Adaptability.* New York: Oxford University Press, 1966.

Bensinger, T. H. "Heat Regulation: Homeostasis of Central Temperture in Man." *Physiological Reviews* 49: 671, 1969.

Cannon, W. B. *The Wisdom of the Body.* New York: Norton and Co., 1967.

Chappell, T., with P. Hackett. "Your Health at High Altitude." *Mariah,* June/July, 45, 1978.

Clements, F. "Some Effects of Different Diets." In Boyden, S. (ed). *The Impact of Civilization on the Biology of Man.* Toronto: University of Toronto Press, 1970.

Cold Injury, Washington, D.C.: Departments of the Army and Air Force, TB MED 81 AFP 160-5-10.

Dubos, R. "Nutritional Ambiguities," *Natural History,* 89(7): 14, 1980.

Frank, J. D. "The Medical Power of Faith." *Human Nature,* August, 40, 1978.

Grey, W. W. *The Living Brain.* London: Penguin, 1961.

Hardy, J. D. "Physiology of Temperature Regulation." *Physiological Reviews* 41: 521, 1961.

Mitchell, H. H. and M. Edman. *Nutrition and Climatic Stress.* Springfield, Ill: Charles C. Thomas, 1951.

Mountain Sickness: Prevention, Recognition and Treatment. Albany, California: Mountain Travel, 1977.

Oyle, I. *The New American Medicine Show.* Santa Cruz: Unity Press, 1979.

Picardi, G. *The Chemical Basis of Medical Climatology.* Springfield, Ill.: Charles C. Thomas, 1962.

Sagan, C. *The Dragons of Eden.* New York: Random House, Inc. 1976.

Tromp, S. W. (ed.) *Progress in Human Biometeorology.* Amsterdam: Swets and Zeitlinger, 1974.

———. *Biometeorology.* Philadelphia: Heyden & Son Ltd., 1980.

Wilson, E., K. Fisher, M. Fuqua. *Principles of Nutrition.* (3rd Ed.) New York: John Wiley and Sons, 1975.

IV

Interpreting Natural Environments

To gain some appreciation of how we have managed to do what no other species had done—namely inhabit the entire earth's surface—we must look briefly at the human brain and its ability to solve environmental problems.

Biological Systems

The Triune Brain

Paul MacLean, a developmental biologist, argues that the human brain evolved in a sequence that strongly favors species survival. The earliest and most primitive functions are concentrated in what today is called the Reptillian Complex (*R-Complex*)—or old brain—the *medulla oblongata* (Fig. 4.1). This is the major control center for the *autonomic nervous system.* It controls homeostasis and the functions of the organs essential to life. To put it another way, you could inactivate some portions of the brain and the body would continue to live as an organism, but without those characteristics we consider to be human. The emotions, group values, memory, thought, etc., are found in sections of the brain that evolved much later than the medulla.

The *limbic system* was the next "layer" of the brain to evolve. Its development began to separate human actions from purely survival behavior. The limbic system generally is associated with emotions, especially emotions such as fear, love, and anger. It is the portion most associated with species (versus individual) survival. It is especially noted for its association with behaviors like *territoriality,* mate selection and protection.

Finally, the *neocortex* and especially the frontal lobes evolved. It is these features that give us our characteristically large head. It is our large brains that makes child birth painful and requires the skull to be open and flexible at birth to permit passage through the female's birth canal. Closure and hardening of the cranium requires about a year and marks a dangerous period that requires prolonged parental (or group) protection. The neocortex is the portion of the brain that controls the least animal-like behaviors, such as

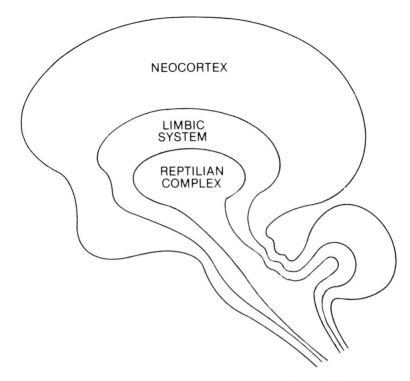

Figure 4.1. A highly simplified representation of the reptilian complex, limbic system, and neocortex in the human brain.

altruism, morality, philosophy, and learning. In short, it is the neocortex that gives man his "human" or "civilized" character. It permits him to rise beyond the survival of self and even of primitive group first behavior.

Given the complexity of the neocortex, it is not surprising that it is the least understood portion of the brain, although we do know some interesting things about it. We know that each side of the brain controls the opposite side of the body. This means that if one side suffers severe damage (such as a stroke) it sometimes is possible to train the opposite side of the brain to assume many of the functions damaged or lost.

We also have recently discovered that the right and left sides of the neocortex, especially the frontal lobes, are associated with distinctly different forms of thought and behavior. The left side is the rational, verbal, mathematical, linear, sequential thinker, the essence of the modern scientific method. The right side is more relational, non-linear and non-verbal. It is more emotional. The right side may also be capable of some elements of precognition, that is, of foreseeing the future. These amazingly different functions and their combinations may go far to explain many of the apparent contrasts between cultures. Contrast between the rational, scientific, logical West and the emotional, metaphysical, philosophical and mystical East may be the result of rewarding or reinforcing different brain functions or characteristics. For example, the people of India exhibit strongly right brain characteristics while European character seems more left brain in behavior. Recently the work of psychologists and others is beginning to show us how to "rationally" make more use of the right side of our brain and how to reward its activity so that we become fuller users and enjoyers of our full mental potential. In a purely technical sense, such a development may give us a much fuller spectrum of abilities to make decisions than we currently believe possible.

Some Possibilities

What we know about the brain and its functional centers permits some interesting speculations. For example, it seems possible that in primitive societies, where survival is based upon skills in hunting and foraging, it is the R-complex that is most active and effective. Other portions of the brain remain seemingly underdeveloped or under-used since their behavior is not crucial to survival. Still, in the primitive context, it could be that the right brain's ability to relate, and its intuitive abilities, are what give the primitive human an ability to survive, despite the fact that other animals often have better physical attributes. What seems evident is that it is only after the basic human needs of food, shelter, and reproductive continuity are satisfied that humans appear ready to indulge in higher intellectual pursuits, such as philosophy and poetry. It is at this stage that the higher evolutionary stages of the brain are used and developed. It makes books and art valuable. It allows us to speculate and deal with environments never personally experienced—to predict and prepare for the future. No other species can do this.

Especially interesting is how the brain's evolution matches the "pyramid of human needs" as presented by Maslow (Fig. 4.2). Note how closely the basic needs are met by the R-complex while the higher states of individual and cultural achievement are elements associated with the Neocortex. It is interesting to speculate on the behavior of not only classes of society but entire cultures as their lives might be viewed in terms of the pyramid. How can a culture or class that must devote its time to providing food and safety be expected to develop the arts or other cultural values?

It would appear that the brain thus contains a built-in priority system. First, the R-Complex must correctly interpret basic biological, survival signals and determine what is necessary for individual homeostasis. The Limbic system provides a family of behaviorally-oriented responses that include such aspects as territoriality and protection of one's immediate family or clan first. It is only the Neocortex and especially the left brain that provides a kind of cultural or moral control that can override the R complex and Limbic system during extreme crises. If this section of the brain is underdeveloped or not used, then humans may revert to earlier stages of development. Such stages place man on a par with any other animal but hardly give him the characteristics we have come to consider human.

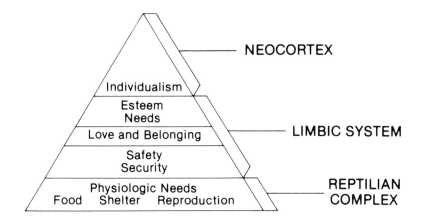

Figure 4.2. Pyramid of human needs according to Maslow. (Data for figure based on Hierarchy of Needs in "A Theory of Human Motivation" in *Motivation and Personality*, 2nd Edition by Abraham H. Maslow. Copyright © 1970 by Abraham H. Maslow. Reprinted by permission of Harper and Row, Publishers, Inc.)

The higher stages of the brain have also provided man with an ability to go beyond mere animal (autonomic) responses in overcoming the environmental stresses inherent in many biomes. Experiences retained in pictographs, petroglyphs, written records and by oral tradition passed on experiences faster than could natural biological imprinting. The ability of the neo-cortex to develop entirely new responses to threat must have continually added new dimensions to human abilities to control or modify environmental surroundings. These abilities are what enabled man to move beyond his strictly biological range of tolerance.

Processing Information

Information and Decision-Making

Given the abilities we have described, how do we humans go about making adjustments to many different environments? Obviously, our minds constantly probe and receive information from our physical surroundings. Proper decisions and actions based on this information allow us to remain in a state of homeostasis within whatever environmental situation we find ourselves in at a given moment. Active probing can be as simple as a brief trip outside on a summer afternoon to see what conditions are like before deciding on the length of a walk. Passive receipt and reaction include things like unconscious unfolding of the extremities in a hot room in response to heat stress. Sometimes direct information is not available, and we must estimate

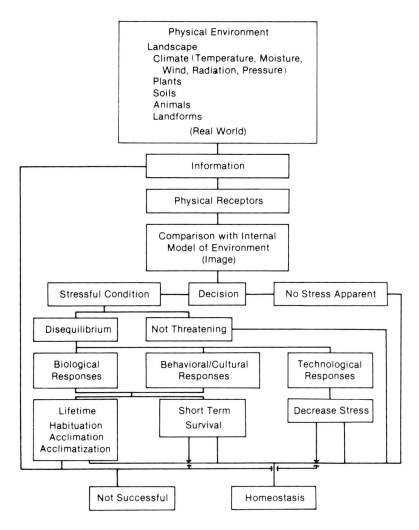

Figure 4.3. A much simplified model of search, reaction and adjustment to environmental conditions

and plan responses before the stresses are actually encountered, like preparation for an expedition to some distant place. The general outline of this search, reaction and decision-making process is shown in Figure 4.3.

The decision-making process begins as we use our natural receptors, the five senses, to receive and collect information from the environment. The head itself (not the brain *per se*) plays two roles here. First, like other extremities, it serves a critical physical sensory role; it contains twenty percent of all the blood in the body and twenty-five percent of the total oxygen supply. The oxygen and blood, of course, are to allow the brain to perform its other function, the reception, interpretation and activation of signals that maintain homeostatic balance and life itself. Although the brain receives myriad messages from the total body surface, there are specific locales which are of paramount importance for reception of thermal signals. The nape of the neck is one critical area of importance for it is here that the most elemental part of the brain is closest to the surface and it is where the massive amount of blood passing to and from the brain is exposed to external temperatures. Other temperature sensors that trigger the homeostatic mechanisms are indicated on the diagram of the body (Fig. 4.4.)

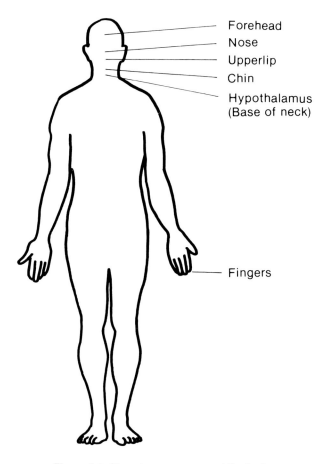

Figure 4.4. Map of sensory areas of the body

The information received is compared by the brain to our existing internal *models* of the environment. Models are made up of a combination of past experience, needs, social customs and other factors. Comparison leads to a decision about the acquired information. If stress is undetected, then action need not be taken because the body is within homeostatic limits.

Perceived stress may lead to disequilibrium, or it may be mild enough so that action is not taken. The coolness of early morning would be an example of a non-threatening stress that passes as the sun warms the ground and air. Stresses that cause disequilibrium must be dealt with by one, or a combination, of three approaches: biological, behavioral, and technological.

Biological and behavioral responses are internal and provide a means for both lifetime and short term reactions to environmental stresses; technological responses usually imply the construction of something. However, technological responses may occur within the energy constraints of a particular natural or cultural system, such as the construction of grass shacks in Monsoon forested areas from native materials. In this instance, the builder is living within the confines of that environment. In some cases, energy subsidies are applied. These may take the form of wood houses built with imported timber upon prairies, or heat added or taken away from dwellings by using imported fuels. Energy subsidies may greatly alter the character of responses to environment and expand the number of possibilities for coping with environmental conditions.

Unsuccessful coping strategies may lead to additional searches for more information from the environment. The number of times unsuccessful search/reaction can occur is limited, however. Eventually the stress would lead to permanent disequilibrium and disappearance of the person or culture unable to make appropriate adjustments.

Finally, some of the elements in the decision process shown on Fig. 4.3 may be less well developed in some individuals than in others. For example, it may be that city dwellers are not as capable of assessing and reacting to natural environments, because actual experience with such environments is limited. It is likely that a city dweller and a person from a society which is involved in a daily tussle with environmental elements for food and water will not construct the same image of a natural environment given the same information. Intead a different picture forms in each person's mind. One image may be constructed of those elements that represent an aesthetically pleasing framework; the other may concentrate on elements that indicate how much food would be available.

Social Behavior

Territoriality

Interpretation and reaction to the environment takes place within a behavioral context called *territoriality*, which seems to be associated with limbic system reactions. There is much argument about the degree and way in which humans exhibit this trait, but there is no doubt that the search and decision-making process is conditioned by spatial perception as well as physical conditions. Recognition of this fact has lead to popular (Ardrey) and academic (Hall, Porteous, Sommer) treatment of territorial behavior among humans.

Territoriality implies exclusive control of a designated space by an individual or group. Exclusivity is reinforced through aggressive defensive measures if the space is invaded by another group or individual from the same species. Competitors for exactly the same resources are eliminated, but other non-competing species may use the space concurrently for their life processes. For example, male mountain lions (*Felix concolor*) claim exclusive territory for their use, but non-competing felines are allowed to exist within the space. Tribal claims among hunter/gatherer societies would be a human analog; some tribes might be principally vegetarians while some may hunt animals. Overlapping territory is used in non-competitive fashion.

Several advantages of territoriality are apparent. Cyclic phenomena like sleep/awake patterns favor a secure place for rest and an equally secure place for the daily practice of food production or work. It is also likely that defense and use of territory led to social organization, communication, and the creation of dominance or leadership structures so that the environment itself and possible intrusions could be dealt with efficiently and effectively. Control of territory sufficient to meet food requirements insures adequate nutrition for the group as a whole; overpopulation is not favored because the environment will not support extravagances. Finally, territoriality conveys a sense of identity for the individual and group, and tells the outside world much about the circumstances you live under, what your thoughts on certain matters are likely to be, and perhaps something about your own level of environmental awareness. Urbanization has made some of these latter attributes less definitive, although territorial behaviors remain but are expressed in other ways. The size of a residential lot, house quality, and property lines are obvious examples.

Space Classification

Porteous has suggested that the space critical to us can be divided into three classes; different search patterns, information, and decision-making occur in each one. *Microspace* or personal space is the first level individuals and groups mark out on the landscape. It is the minimum space needed for a normal existence. It is movable as the individual or group changes positions, and it is defended when encroached upon. The individual, family, or house serve as examples under this category, although many other groupings are possible. Microspace boundaries are very clear here: a child's bedroom, a parents' office, the family home, my side of a dormitory room, Fred's favorite chair.

Outside of microspace lies *mesospace,* our home base, or neighborhood, the area we consider essential for our daily living patterns. It, like microspace, is movable as the individuals and groups we associate with move about. Micro and mesospace constitute the range of environmental conditions we are most familiar with and are the information base that we use to test and react to unfamiliar environments.

Finally, *macrospace* encompasses our total range within the territory we or our group has staked out as ours. We use macrospace for acquiring food from units as small as the grocery store or as large as the grasslands of the plains. Macrospace is also used to satisfy other physical and psychological needs. Other groups may use the same environment for similar purposes at the same or different times. This is public space where identity as an individual or group is not necessary but identity as a whole, like states or the United States, takes place. We do not know much about all of the nuances of the macrospace environment because we are not continuous occupiers of it. Macrospace requires us to expand our catalog of mental information and to bring interpretive skills to bear on those unfamiliar aspects of environment that seem important.

We turn now to a more detailed description of the spaces and their characteristics which humans occupy.

Terms

Altruism. Regard and devotion to the interests of others.

Autonomic Nervous System. That part of the peripheral nervous system which regulates involuntary responses, especially those concerned with nutritive, vascular, and reproductive activities. Often called the R-Complex.

Limbic System. The second section of the brain to evolve; basic emotions, such as fear, anger, are centered here.

Macrospace. Public space occupied by us at times and by others at the same or different times; used for securing food, other needs; not movable.

Mesospace. An individual's neighborhood, the space that provides most of our knowledge about the physical environment; not movable.

Microspace. Our immediate surroundings, the minimum space needed for survival; movable space concept.

Model. A general, often hypothetical, description or representation of something that can be used to examine, explain, or analyze the object.

Neocortex. The last section of the human brain to evolve; basic attitudes, motivations, sense of humor, abstract thinking arise in this section of the brain.

R-Complex. The most primitive part of the human brain, thought to have evolved first, and to be associated with autonomic responses.

Territoriality. Declaration in various ways that a specific area is under the control and is for the use of a particular individual or group. An adaptive mechanism that allows construction of a mental catalog of environmental information and coping strategies.

References

Ardrey, R. *The Territorial Imperative.* New York: Atheneum, 1966.

Downs, R. and D. Stea. *Maps in Minds.* New York: Harper and Row, 1977.

Hall, E. *The Hidden Dimension.* Garden City, New York: Doubleday, 1966.

Holden, C. "Paul MacLean and the Triune Brain." *Science* 204: 1066, 1979.

Loye, D. "The Forecasting Mind." *The Futurist,* June, 177, 1979.

Porteous, J. D. *Environment and Behaviour.* Reading, Mass.: Addison-Wesley, 1977.

Sommer, R. *Personal Space.* Englewood Cliffs, N.J.: Prentice-Hall, 1969.

Saarinen, T. *Environmental Planning—Perception and Behaviour.* Boston: Houghton-Mifflin, 1976.

V
Natural Environments

Human activity occurs within a broad range of environmental conditions, and different conditions require different adaptations and adjustments. By emphasizing the biome concept as a way of summarizing important environmental features of different environments, we can produce a framework for looking at human adjustments to various physical environments.

Environmental Structure

Biome Structure and Function

All natural environments consist of an interlocking mosaic of elements similar to those shown on Figure 5.1. Each box in the figure comprises a set of physical and biological elements. Their interaction creates a biome, or an assemblage of living and non-living features characteristic of natural environments. Biomes are defined as energy-based systems for our purposes. They are characterized by a cycling of materials (nutrients), energy flow, *self-regulating* (feedback) *mechanisms, uni-directional energy transfer,* and *hierarchical differentiation,* an organizational framework that has step-like features in it. Acting together, these features balance incoming energy (primarily sunlight) and its disposition. The amount of incoming solar energy *(insolation)* varies widely from place to place, as does the way it is disposed of in a biome.

Man can either interact within a biome as though he were an integral part of it, and, thus be in balance yet subject to internal energy relationships, or he can act as a disruptive user of energy resources. Such a role might not only upset an individual biome, but may also destroy its ability to support human life.

Feedback mechanisms serve to keep a biome in energy equilibrium in much the same way that homeostasis maintains a constant temperature in the human body. The more energy a biome receives, the greater the diversity of plant and animal species it can support. Availability of many food sources make the system more difficult to upset or destroy permanently. Conversely, the more limited energy and food are (a measure of severity), the fewer feedback mechanisms there are, and the more difficult it is to maintain biome homeostasis.

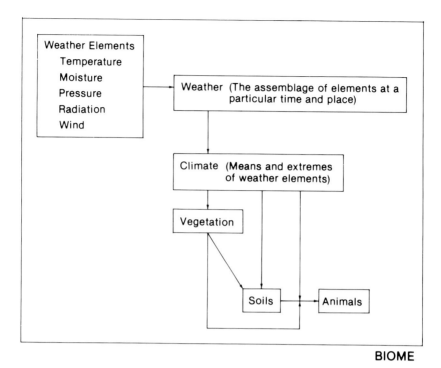

Figure 5.1. Interaction of physical and biological factors in natural environments

Feedback processes are often described in cybernetic terminology. For example, let us assume that the number of deer in a severe (energy limited) biome, such as the tundra should increase because of an unusual increase in the food supply. This increase in the deer population would be offset by an increase in *predation* that would help keep deer numbers in balance with normally available energy. Thus, the only way to permanently increase the number of deer would be to permanently increase the food supply. The number balance we are speaking of is maintained by *negative feedback,* or feedback that constantly works to promote stability or maintain the *status quo.*

If negative feedback fails, then *positive feedback* may occur. An example of positive feedback using deer might be overharvesting. Overhunting would lead to a smaller breeding population, which leads to fewer surviving young and a further decrease in herd size.

Figure 5.2 illustrates both energy transfer and hierarchical differentiation within biomes. They are important because they control the way in which energy flows through a biome, and they allow us to describe the amount and position of available energy for plant, animal or human use at any time.

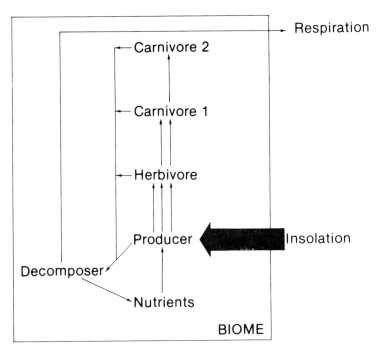

Figure 5.2. Structure and function, energy flows, and nutrient cycling in biomes

The initial source of energy for all biomes is the sun. Plants *(producers)* use only a small part of available insolation, and this they convert into chemical energy which in turn is locked up in plants (biomass). A small portion of this energy is then passed to *herbivores,* or plant-eaters. Herbivores, in turn, are preyed upon by *carnivores* (meat-eaters); some animals (omnivores) consume both plants and animals for food. Some energy is lost also to *decomposers,* and some is lost through *respiration* as living things go through life cycles. In all cases, energy is transferred from a level where it is more abundant to another. Each level or step is called a *trophic level.* All trophic levels are linked in a *food chain,* a ladder-like arrangement. The number of steps in a chain is limited by the amount of plant production. Chains may be linked together in *food webs:* animals or plants from one food chain may also use food sources from another. The number of chains and webs can tell us much about the adaptations and adjustments humans must make in different biomes.

Productivity

The number of chains, webs, and trophic levels reflects the *productivity* of a biome. Productivity in unmodified biomes is controlled by environmental variables. Humans have been able to change the normal characteristics of nutrient cycling, plants, water availability, or energy exchanges in an area and have greatly increased productivity in some areas. Careless modifications have led to the opposite situation.

The productivity of different biomes can be compared by harvesting *biomass,* or standing crop, and multiplying it by an energy equivalency factor that represents an estimate of the amount of energy produced per square centimeter by native plants over time. The result of one such exercise is shown on Fig. 5.3.

It is apparent that different biomes vary greatly in productive capacity. We could also say that the amount of available energy in native plant material, which forms the base of any food chain, is tremendously different from place to place. The least productive biomes are found at high latitudes or occur in areas with little rainfall. Conversely, the most productive areas lie near the equator where insolation is greatest. For example, equatorial areas comprise only about eleven percent of the earth's solid surface, but they account for one third of all the energy in plant material.

The biome concept allows us to estimate the needs and ease with which man adapts and adjusts to the natural environment. Clearly, in some areas there are abundant energy and diversity of food sources naturally available to permit human occupance with only minimal technological adjustments to offset environmental stress. Other biomes provide not only thermal stress (which we looked at in Chapter Three) but food and energy stress as well. Thus, we could define a severe biome in terms of human use as one in which human food is only minimally available. Stated another way: environmental severity is determined by how much of your natural environment you cannot eat.

Biome Concept and Human Strategies

It should be clear at this point that each biome in its natural state has a *carrying capacity* which is linked to productivity. If humans are to occupy successfully any natural environment, we must either adapt to the biome as merely one more species in the balanced system, or we must employ a variety of strategies that enable us to overcome or offset the natural limits of the environment.

In the process of adapting to environmental stress and circumventing natural limits posed by the environment, humans have assumed a dominant position among myriad species. Making proper choices from among a variety of options—including migration away from and minimal occupance of inhospitable environments—has been continuous. Poor choices ended in disasters. Successful adaptations were quickly copied and often institutionalized. In all instances each adaptation required some awareness and adjustment to the natural energy balance of the biome itself.

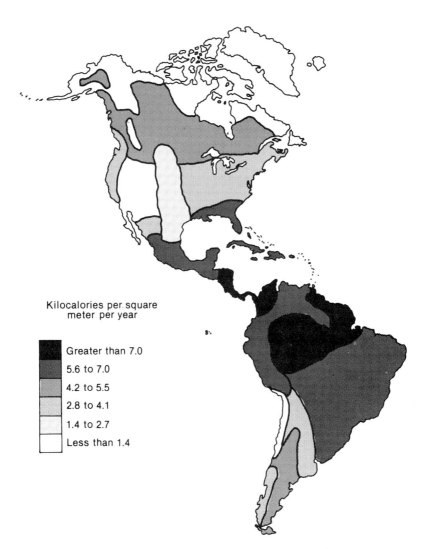

Figure 5.3. Generalized estimate of biomass productivity. (Adapted From H. Lieth, 1975. Primary production of the major vegetation units of the world, p. 208 in H. Lieth and R. Whittaker, eds. Primary Productivity of the Biosphere. New York: Springer-Verlag.)

A. No Energy Subsidy

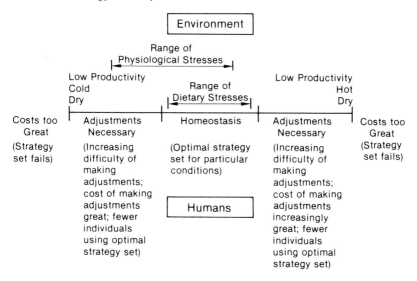

B. Energy Subsidy

Figure 5.4. Stresses and adjustments in hypothetical natural and energy-subsidized environments

Figure 5.4a illustrates the range of coping strategies in hypothetical natural and energy-subsidized environments. The top of the diagram represents man living in an unmodified biome without any artificially supplied energy. Testing and adoption of various strategies, such as diet, housing, clothing, etc., assures survival of at least some people under even highly stressful environments, like the arctic tundra or deserts. Each biome requires some combination of such strategies if we are successfully to occupy that environment or location. The harsher or more severe the environment, the more limited the choices, the richer the environment (particularly in food supplies), the wider the range of adaptive behavior and adjustments.

Living within energy constraints is the way all humans once dealt with earth environments and is still the approach in some regions. We are quite capable of existence within the established food chain structure present in a particular biome, and compete successfully with elements in the food chain for some of the available energy. This is the most deterministic relationship we have with environments because survival strategies must be geared to existing conditions.

Clearly, physiological need, mobility, and intelligence also are involved in determining the range of choices in natural biomes. Humans choose a set of adaptive strategies that assures maximum success under given environmental conditions. When these conditions change, a behavioral strategy may no longer prove effective. People either adopt or develop new strategies, or they continue to use those they know work and return to a biome where they are effective. If environmental factors ultimately prove too difficult to overcome, an individual or group will not be able to successfully live in that particular environment. A strategy like hunting and gathering would prove useful in dissimilar biomes, but other adaptations, such as clothing patterns, would not be equally transferrable.

With energy subsidies, such as fire and the use of fossil fuels, a whole new range of possibilities emerged (Fig. 5.4b). A new dimension to the decision-making process was added and has led to prodigious food production on often marginally productive land. Today, shipping and trucking permit large interregional shifts of foodstuffs that may produce a similar diet for many people in the United States and many industrialized countries regardless of the natural biome and its constraints. We have even eliminated the need for any food production in some biomes, like that of the cold desert in the United States. Food strategies are non-existent in this case, except for a trip to the market. Where traditional food gathering activities still prevail, energy subsidies like guns and snowmobiles, improve the efficiency of the hunt.

It is also possible to create a more constant set of optimal conditions through energy additions; clothing and housing serve as examples.

Major Biome Types and Energy Relationships

Strategies for human adaptation must confront conditions found in actual biomes. Figure 5.5 shows four principal global biome types; Forest, Desert, Tundra, Grassland, on a map that also shows subtypes within each of the four types. General environmental characteristics in each biome are shown on Figure 5.6.

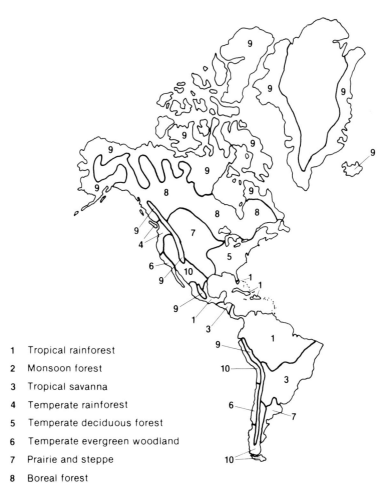

1 Tropical rainforest
2 Monsoon forest
3 Tropical savanna
4 Temperate rainforest
5 Temperate deciduous forest
6 Temperate evergreen woodland
7 Prairie and steppe
8 Boreal forest
9 Arctic and alpine tundra
10 Deserts

Figure 5.5. Generalized model assemblages of climate, vegetation, soils, and animals that interact together to form a biome.

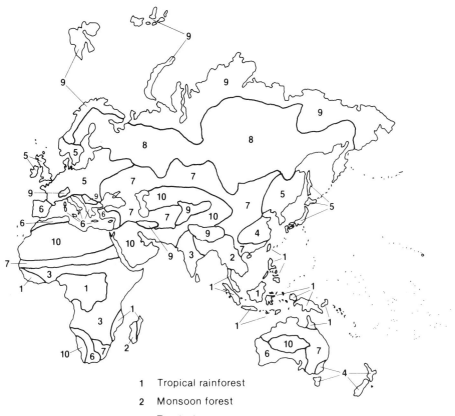

1   Tropical rainforest
2   Monsoon forest
3   Tropical savanna
4   Temperate rainforest
5   Temperate deciduous forest
6   Temperate evergreen woodland
7   Prairie and steppe
8   Boreal forest
9   Arctic and alpine tundra
10 Deserts

Figure 5.5—*Continued*

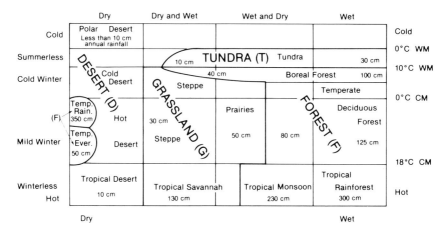

Figure 5.6. Simplified position and environmental conditions of model biomes arrayed on a hypothetical continent.

Table 5.1. Characteristics of Selected Biomes

Name	Characteristics
Tundra (T)*	Long periods of Darkness Followed by Long Days
	Low Biological Productivity / Short Growing Season / Acidic Soils
	Very Low Average Temperatures for Long Periods
	Little Surface Moisture / Most in Frozen Form
	Permafrost
High Altitude Tundra (T)	Little Oxygen Available
	Great Diurnal Temperature Range / Low Average Nighttime Temperatures
	Low Biological Productivity / Short Growing Season / Acidic Soils
Desert (D)	Low Biological Productivity / Insufficient Moisture / Low Organic Matter in Soils
	High Average Daytime Temperatures / Large Diurnal Range
	Extreme Evaporative Stress on Species
	Low Average Rainfall / High Variability in Amounts Received
	High Solar Radiation Loads
Prairie / Steppe (G)	Low Rainfall Totals / Periodic Drought
	Large Diurnal / Seasonal Temperature Range
	High Evaporative Stress in Summer
	Seasonality of Vegetative Growth
	Fertile Soils

Table 5.1—*Continued*

Name	Characteristics
Tropical Forests (F)	Rapid Mineral Cycling High Rainfall Totals High Average Daytime Temperatures Cooling Stresses Pronounced Because of High Humidity Soils Impossible to Crop Continuously High Productivity but only Native Flora/Fauna
Boreal Coniferous Forest (F)	Short Growing Season Large Seasonal Changes in Average Temperature Long Cold Periods Low Moisture Totals Highly Acidic Soils Intermittent Permafrost
Temperate Deciduous Forest (F)	Acidic Soils of Moderate Fertility Relatively Short Growing Season Moderately Low Avg. Annual Temp.
Temperate Rainforest (F)	High Rainfall During Period of Low Evapotranspirational Stress High Runoff Percentage Low Light Intensities Infertile Soils/Slow Cycling of Nutrients Food Chains not Complex or Represented by Large Numbers
Temperate Evergreen Woodland (F)	Large Seasonal Differences in Precipitation Amounts Infrequent but intense Droughts High Moisture Stress During Summer Limited Productivity Limited Food Chains and Numbers Slow Mineral Cycling

*T tundra
D desert
G grassland
F forest

Specific characteristics that must be dealt with by humans in major biomes are contained in Table 5.1, and are summarized on Fig. 5.7. For example, in Tundra areas, humans face problems associated with low biomass productivity, extreme cold, and restricted dietary resources. Such a combination of limitations favored a human pattern of occupance marked by settlement along seacoasts, fishing, high fat intake, layered clothing and housing that takes advantage of human-produced (anthropogenic) heat. Conversely, in Tropical Rainforest areas, heat stress is a problem, but there is a wider choice of food stuffs growing on marginal soils. Thus, this biome has favored

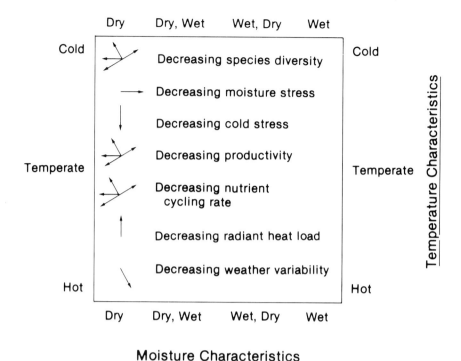

Figure 5.7. Some changes in biome characteristics with changing environmental conditions.

a pattern of activity that includes minimal gathering, low fat intake, minimal clothing and shelter. Each of these biomes has low population densities, Tundra because of low biomass productivity, and Rainforest because usable human food sources are limited. Intense agriculture is not possible in either region due to the nature of the physical environment and, without huge energy subsidies, a large human population would quickly deplete the available food resources in either area.

The remainder of this chapter comprises an illustrative but not exhaustive look at human adjustments in a number of major biomes, the limits they pose to human occupance and a look at some successful strategies in their use by man.

Biome Characteristics

Tropical Biomes

The Tropical or low latitude series of biomes contains examples of three of the four principal biome classes. In this group, we find that thermal and evaporative stresses are high, radiation intense, and productivity variable from low (Savanna, Deserts) to high (Rainforest). Population densities are low, except for urbanized areas and in particularly favorable locations associated with the fertile alluvial soils along river courses or near water sources.

The *Tropical Rainforest* Biome comprises about eleven percent of the land area of the earth. It is characterized by high average annual temperature, high insolation, high humidity and abundant moisture throughout the year. A great number and variety of food chains and webs, rapid nutrient cycling, and the highest natural biomass productivity of any biome are also characteristic.

Natural vegetation and undisturbed soils are still predominant in this biome, and animal populations are composed of types and numbers similar to those present before human occupation. Most of the biome's nutrients are in the standing crop, not in the soil. This is a major limitation to sedentary agriculture. Also, soils cannot be fertilized using mid-latitude techniques, techniques that work in other biomes. They are highly susceptible to erosion, and may become extremely hard *(laterized)* when the vegetation cover is permanently removed.

Natural vegetation is characteristically multi-layered and there are many species. The vegetation is mostly evergreen and often creates a closed cover that limits light penetration to the forest floor (Fig. 5.8). Vines and plants that use other plants for support and nutrients are common and most animal species are aboreal.

The forested landscape produces a great number of kinds of foodstuffs—from nuts (e.g., cashews) to fruits (papaya, mango, banana), leaves, shoots, tubers, small animals—but not large numbers of any one kind within a small area. Food sources are spread over a considerable area, and both animals and man must traverse some distance in the gathering process.

Agriculture is practiced using slash and burn techniques. This is a practice that entails the slashing of trees to kill them, followed by a burning of the dead vegetation to both clear the ground and add nutrients *(fertilizers)* from

Figure 5.8. Tropical rainforest biome overview. (Courtesy of W. Johnson.)

the ash (Fig. 5.9). Burning allows the sun to penetrate to the forest floor. By breaking down organic matter, burning also transfers some nutrients from the standing crop to the soil. But once the ground is exposed to the sun and rain as well as being tilled, the soil quickly becomes exhausted and the area must be abandoned so that it can recover. Meanwhile, a new area has been cleared in a kind of land rotation.

The Tropical Rainforest biome is, thus, a study in contrasts. It looks and is vegetationally lush, diverse, and productive in biomass. Unfortunately, the system is delicately balanced and cannot be made to produce much more, particularly by farming. In order to take advantage of the biome's resources, people must move from place to place. This is a life style that can only support low population numbers and densities.

Monsoon Forest Biomes are characterized by a strong wet/dry cycle. They have a five to nine month rainy season, during which rainfall totals may exceed those in adjacent rainforest areas, and a three to seven month dry period; rainfall totals for these latter months are often less than fifty cm.

Figure 5.9. Slash and burn agriculture.

Figure 5.10. Rice fields in a monsoon forest biome.

The monsoon biome occupies about five percent of the earth's land area; its productivity percentage is about ten. Vegetation in this biome is not as tall as in Tropical Rainforests. It often is found in fewer layers, and is a mixture of deciduous as well as evergreen species.

Sedentary agriculture in river valleys on alluvial soils is a common practice. Soils away from the rivers and deltas are similar to other tropical soils, but they are slightly more fertile. Animal populations are less diverse than in tropical rain forest areas and are present in fewer numbers. Energy for farming is provided by humans and large hervivores. Rice typically is the staple crop; native populations utilize the lowest level on the food chain (Fig. 5.10).

Tropical Savanna is a third type of tropical biome. Land area and productivity percentage figures are about ten and twelve, respectively. The climate is characterized by strongly seasonal changes in the amount of solar heating. Rainfall during the three to six month period of high sun exceeds fifty cm. It falls substantially below this figure during drier periods. The result of these cycles is that grasses tend to be the dominant form of vegetation, although small trees may be interspersed at frequent intervals. Soil fertility varies widely, although most soils tend toward low fertility. Extensive populations of herbivores such as zebras typically roam the tropical savanna. Large carnivores, like the African lion, prey upon herbivores in a well-defined food chain. Human efforts to produce food often parallel the natural systems. Agriculture typically is marked by grazing of domesticated herbivores (cattle). Sedentary crop agriculture is limited to areas with reliable supplies of water. In wetter portions, those with a relatively long rainy season, bush-fallow farming is common. This is a technique similar to the rainforest slash and burn practices. It involves tree cutting, stacking of the wood, burning to return nutrients to the soil, and cropping with grains, sugarcane, cotton and other crops. Soils often are exhausted after several seasons and must be allowed to recover. This again means that rotation must be practiced.

Temperate Biomes

A second major biome group consists of the Temperate or Mid-Latitude series of biomes. With temperatures averaging 10°C (50°F), overall thermal stress on the human body is at a minimum. However, as we shall see, certain elements may create stress at various seasons. It is in the temperate biomes that we find the densest concentration of humans.

The *Temperate Rainforest* biomes are concentrated along the west coasts of continents and are marked by a cycle of cool summers, with little rainfall but much fog and cloudy conditions, and winters that are cool to cold and are accompanied by substantial rainfall. During winter, vegetative growth is minimal and, therefore, transpiration (water loss through plant leaves) is minimal; extensive runoff during the wet season is the result.

Dominant vegetation is tall coniferous trees that effectively block light from the forest floor much as in the Tropical Rainforest. As in the rainforest, this limits plant productivity below the canopy.

Soils are generally acidic and prone to erosion if the vegetation is removed. They are not exceptionally fertile.

Because of the milder temperatures and abundant moisture in the Temperate Rainforest, species diversity is higher here than in other temperate forest biomes. This means that food webs are present. But they are not as complex as in the tropical forests. Typically, elk and muledeer are the major herbivores and mountain lions and wolves act as the major carnivores.

Figure 5.11. Temperate rainforest biome.

Due to seasonally cool temperatures, decomposition of organic matter (humus) is slower than in tropical rainforests. Plant nutrients are not readily available as a result. Most of the biomass production goes into the growth of large trees. This in turn limits food for the herbivores and animals higher on the food chain (Fig. 5.11).

Agriculture is not widely prevalent. It is confined largely to clearings where sunlight is available. Populations that occupy the Temperature Rainforest typically have used fish as their major source of food. Logging of Douglas fir and Redwood trees, which often reach heights in excess of seventy meters, predominates as a non-agriculture economic activity in coastal areas.

Temperate Evergreen Woodland also is found along the west coast of continents, but at lower latitudes than temperature rainforests. A typical example is found around the Mediterranean Basin. The area occupied and this biome's biomass production are about three and five percent of the solid earth's surface and terrestrial productivity, respectively. The major environmental cycle is associated with large latitudinal shifts in the westerly wind belts brought about by seasonal changes in intensity of the Subtropical High pressure cells. High pressure, high temperatures, and clear skies are accompanied by great water stress associated with only minimal rainfall during the dry season (April to September). During the wet season, rain from passing cyclonic systems is common. The cool season extends from November to

Figure 5.12. Temperate evergreen woodland biome.

March and is accompanied by rainfall, which often amounts to as much as ninety-five percent of the annual amount. Unlike the rainforest, soils are alkaline. However, these soils are also low in organic matter, but fertile if moisture is available. Vegetation is composed of short, woody, mostly thick-leaved evergreen plants that vary in spacing from impenetrable to widely dispersed. Grass occupies the openings between woody species (Fig. 5.12). Productivity is low primarily because of the long, dry summer season; rapid mineral cycling for plant growth and large populations of herbivores and carnivores are not favored by such a regime. Species diversity appears to lie about halfway between grassland and temperate deciduous forested areas. Prehistoric dominant animals included the grizzly, pronghorn antelope, deer and elk. Today herbivores remain as dominants.

Agricultural practices vary widely from place to place within the biome, but similarities are evident. Grapes, orchards, and small grains are common today in all areas; extensive grazing also is practiced in Australia and the valleys along the U.S. west coast. Because the pleasant climatic circumstances are so attractive to man, much of the original character of this biome has been changed over the centuries, particularly around the Mediterranean Basin.

Figure 5.13. Temperate deciduous biome.

Temperate Deciduous Forest biomes are regions of high population density and industrial activity. In this biome, climate is controlled by the global Westerlies. Large moving cyclonic storms bring winter rainfall; summer conditions are dominated by warm, moist air invasion from oceans. Precipitation varies from high values (120 cm) in the east to less than seventy five cm along the western border and is fairly evenly spread throughout the year with a bias toward the summer months. Temperatures can reach very low levels at the poleward limits of the biome, at times even lower than those of Arctic or Antarctic regions. High temperatures and high humidities prevail in summer; the region thus is characterized by a distinctly hot-cold cycle of seasons. The biome reaches maximum development in North America in the Great Smoky Mountain region in the United States. Here trees reach maximum size and the multi-layered understory reaches maximum complexity when compared to other locations (Fig. 5.13).

Animal species commonly found throughout the biome include the white-tailed deer, turkey, fox, squirrel, skunk, and, in the last century, the wolf and black bear. Many animals show marked seasonality, with hibernation common in winter as an adaptation against cold stress and lack of food sources. Food chains are well developed and food webs are common in relatively undisturbed areas, although extensive modification has curbed the diversity and numbers of animal species present. The northern limit of the forest is marked by cold conditions and a transition to the Boreal Forest Biome; to the west, moisture proves limiting until only drought tolerant species appear along river valleys in the Prairie and Steppe biomes.

Soils are thin, acidic but reasonably fertile under a typical agricultural regime that includes rotation of different crops on the same field from one year to the next. Agriculture varies from cash grain crops in the western portion to truck gardening in the east and cotton in the south.

Prairie and Steppe lands comprise some six percent of the earth's surface and account for about five percent of terrestrial productivity. The vegetation is predominantly grass. On its wetter margins, the grass is tall and interspersed with forests. Along its more arid margins, the grasses are short, but with cottonwoods in very moist areas such as river beds (Fig. 5.14). The Prairie biome (tall grass) occupies regions where precipitation exceeds evaporation; steppe (short grass) also is found in drier areas where evaporation exceeds precipitation.

The climate of these biomes is marked by great shifts between summer—usually very hot—and winter—markedly colder, e.g., summer average maxima are 26°C (79°F) while winter averages −3°C (28°F). This climatic cycle is created by changes in dominance of cold, dry (winter) and warm, moist (summer) air masses which in turn are manifestations of the general circulation of the atmosphere as it responds to annual changes in solar radiation. As one would expect, moisture stress is high during the winter months

Figure 5.14. General view of a prairie environment.

given the dryness of the air and the frequent presence of relatively high winds. Summer is marked by similar stresses, which may become severe if drought, a frequent visitor to such areas, sets in and persists.

Grassland biomes are grain producing areas. Thick soils, rich in organic material and nutrients, provide a firm basis for extensive agriculture. Drought can modify productivity figures substantially, as recent and recurring problems in the Soviet steppes illustrate. Relatively simple food chains composed of large numbers of similar plants and herbivores once prevailed across the North American grassland, but pressure for agriculture has reduced the great herds that once roamed the region. In the Great Plains of the U.S., bison and pronghorn antelope dominated as herbivores; only the pronghorn remains in significant numbers. Carnivores include coyotes, weasels, badgers, foxes, owls, and rattlesnakes, all small and mostly nocturnal to avoid high daytime temperatures. Burrowing is common for insulation from winter cold.

Cold Biomes

High latitude (Arctic areas) and/or high altitude (mountain) biomes comprise a third group of biomes. Cold stresses combined with large seasonal contrasts in insolation, low biomass productivity, short growing seasons, lack of species diversity, and simplicity characterize this group.

Figure 5.15. Boreal forest biome. (Courtesy of D. Butler.)

The *Boreal Forest Biome* covers approximately eight percent of the solid earth surface and accounts for seven percent of its NPP, principally because of its tremendous size (Fig. 5.15). It is an extreme environment with several months averaging above 10°C, many months below 0°C, and a very large range in average temperatures (some 60°C over Siberia). Moisture falls mostly as snow. The growing season is only about sixty days but the daylight of this period is very long.

Soils are very high in near surface organic matter because low temperatures do not allow rapid decomposition. The soils are acidic and not very fertile. Vegetation is composed of predominantly two coniferous tree species, both spruce, an indication of low diversity. There are few food chains, webs, and numbers of animals. The region is a continentally glaciated landscape and is, therefore, flat. Flatness often means poor drainage and swampy conditions during the warmer season when the frozen ground thaws.

Agriculture is a difficult endeavor and is practiced at risk. Some cereals and vegetables can mature in the long days of summer, but the short growing season places restrictions on the variety of crops that can be grown. For these and other reasons, the region is not densely populated. The cities of large size found within its confines occur principally in the Soviet Union or coastal areas where commercial fishing predominates. Dominant animals include moose, caribou, elk, grizzly, beaver and several species of birds.

The *Arctic Tundra* is a vast treeless biome region characterized by severe winters and somewhat less severe summers. The growing season is very short although daylength is long. Radiation totals are high during the summer, yet temperatures may fall below freezing any month of the year. Little precipitation is received at daytime. The area is underlain by permafrost; most solar radiation during summer melts the upper layers of soil leaving little energy for heating of the air. The natural vegetation is composed of low-laying grasses and sedges. Accumulation of organic matter at the surface prevails because temperatures are not high enough to allow rapid breakdown of plant materials. Food chains and webs are simple and composed of few species but large numbers. Brown bear, reindeer, caribou, musk ox and wolf are common large animals; hares, owls, lemmings, and waterfowl are also common.

Arctic Tundra shares many of the same characteristics of *Alpine Tundra*. In alpine areas, elevation replaces latitude in creating similar environments at a variety of locations (Fig. 5.16). Alpine areas have plant species common to Arctic areas, and often are as severe climatically. However, high elevations have higher radiation loads than their arctic counterparts because of the thin atmosphere at such levels, and they do not always share the long season of light.

Figure 5.16. Tundra landscape.

Deserts

Desert biomes occur at low, middle as well as at relatively high latitudes. They typically are divided into cold and warm deserts. Warm deserts are found in the latitudes nearest the equator, in the interior of continents and along the western coasts (such as Southern and Baja California). Higher latitude as well as high altitude deserts are much colder than their equatorward counterparts. They lie in the lee of mountain ranges or are far from moisture sources. Examples include Central Asia and Central Canada. Together the warm and cold deserts comprise some twelve percent of the earth's surface but account for only one percent of its productivity.

By definition, rainfall is minimal. Due to the lack of moisture, skies are normally clear and insolation is intense. Species diversity is limited and the food chains and food webs are only poorly developed. Both large animals and large plants are uncommon. Due to competition for limited water, vegetation is widely dispersed and made up of species with morphological and physiological adaptations to extreme moisture stress (Fig. 5.17).

Soils are alkaline and poor in organic matter cycles. Native peoples, like the Kung in Africa, provide their foodstuffs through hunting and gathering. However, the number and kind of edible items is distinctly limited, and natives must move from place to place because food supplies are quickly exhausted. This means low population numbers and few fixed settlements. Some grazing activity does occur but sedentary agriculture is limited to intensively irrigated regions, such as Central Asia, Egypt, where population densities may become very large, but limited in expansion by the desert's margins and its physical limits.

Figure 5.17. Desert biome landscape.

Terms

Biomass. The amount of living material in an area or volume.

Carrying Capacity. The amount of living material supportable by physical elements in a biome.

Carnivore. Flesh-eating mammal that consumes herbivores in order to supply itself with energy for metabolism and motion.

Climate. Means and extremes of the five weather elements.

Decomposer. Typically a soil-based organism that breaks down complex molecules that constitute dead producers, carnivores and omnivores thereby providing nutrients which can be reused by producers.

Fertilizer. Nutrients added to soils in liquid or solid form. Nitrogen, phosphorous, and potassium are called macronutrients because they are needed in large amounts by plants. Micronutrients are essential for growth and are supplied in small amounts where soils are naturally deficient.

Food Chain. A way of describing where different biological elements in a biome derive energy for metabolism and growth. Always starts from a producer base and may continue through herbivores or omnivores to a carnivore stage. At each successive consumptive stage or step, called a trophic level, energy becomes less available. Thus, the number of steps and the number of individuals that can be supported in a biome are limited by the productivity of the producer segment.

Food Web. Interlocking food chains. One animal might take energy from two different food chains. The number of webs increases with the productivity of a biome.

Herbivore. A plant-eating consumer. Size may range from large ungulates like moose to small ones, such as the rabbit.

Hierarchical Differentiation. An organizational structure that divides biome elements by numbers or available energy.

Insolation. Direct plus diffuse (scattered) solar radiation that reaches the earth's surface.

Laterized. A soil with very high amounts of aluminum and iron in its upper layers. Many other elements have been removed from upper soil layers by water action. The resulting soil becomes rock hard if exposed to sunlight for long periods.

Negative Feedback. A type of self-regulating mechanism that works against change thereby promoting homeostasis or equilibrium.

Positive Feedback. A type of self-regulating mechanism that works to promote change and thereby departure from homeostasis or equilibrium.

Predation. The act of hunting to secure food.

Producer. The lowest level on a food chain, namely plants.

Productivity. A measure of the total yield of biomass in a biome. Often the term is restricted to plant (producer) productivity and is measured by harvesting plant biomass or standing crop.

Respiration. Process of oxygen intake by a living organism and use in oxidation of nutrients thus providing energy for metabolism.

Self-Regulating Mechanisms. Feedback mechanisms that assure homeostasis unless changed.

Trophic Levels. Energy levels in a food chain. Each level represents the total energy available at that point.

Uni-Directional Energy Flow. Flow from a source to a sink, from places or levels of abundance to places or levels of scarcity. Operates in a biome; energy passes from the sun through producers to consumers.

Weather Elements. Wind, radiation, temperature, moisture, pressure that represent climate when summarized by means and extremes.

References

Billings, W.D. *Plants and the Ecosystem.* Belmont, CA: Wadsworth Co., 1978.

Clapham, W. *Natural Ecosystems.* New York: Macmillan Co., 1973.

Shelford, V. *The Ecology of North America.* Urbana: University of Illinois Press, 1963.

Watt, D. *Principles of Biogeography.* London: McGraw-Hill Co., 1971.

VI

Adaptive Responses to Biomes: Clothing and Housing

Once we understand the basics of the human organism and the various environments in which it must survive, we have set the basis for understanding how humans use and respond to terrestrial biomes. The key factors in understanding how man has been able to occupy so many diverse biomes, biomes well beyond his apparent biological limits are, first, his mobility, and second, his adaptability.

Clothing and Housing as Responses to Environment

The assessment and the decision-making processes have led to adjustments and adaptations that are remarkably similar in function, but perhaps not form, under similar biome conditions around the globe. Clothing and housing are two mechanisms humans have learned to use against less than ideal environments. Few features have been more important in permitting the peopling of the earth. Both are similar because both mitigate against stress by altering energy exchanges between the body and the environment. We use both to insulate the body from heat gain or heat loss and we limit or encourage moisture loss (perspiration) due to conduction, convection, advection, moisture, and radiation.[1] Decision factors that apply to both are summarized on Figure 6.1.

In the following three sections, we look at clothing and housing. The first two describe functional characteristics of each which are considered in their design to make them effective adaptations to environmental conditions. The third section illustrates clothing and housing practices in the biomes described in Chapter Five.

1. Terms are defined in previous chapters.

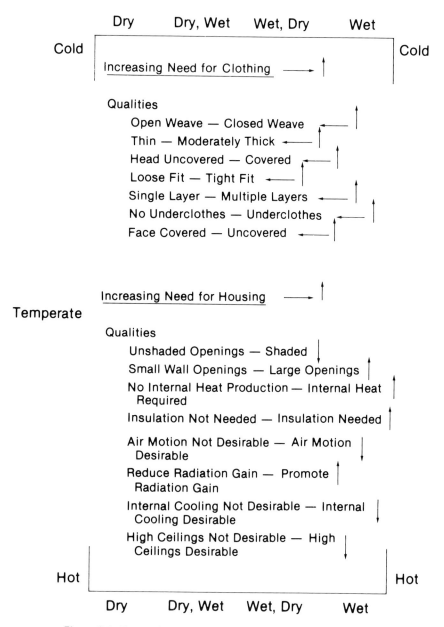

Figure 6.1. Factors important for clothing and housing design under different environmental conditions.

Clothing

Clothing represents a personal way of dealing with whatever stress the environment has to offer. Regardless of variations in style or cultural preference, we all face the same clothing problems.[2] These common problems are largely determined by the environmental conditions we reviewed in Chapter Four.

In some biomes, clothing problems are minor. For example, in hot/wet conditions (Tropical Rainforest), clothing is limited largely to what modesty dictates. The ideal adjustment is to leave the body a free surface that can conduct, convect, radiate, and evaporate as much heat as possible; clothing would merely restrict this natural cooling. Under more severe conditions, the problem of clothing becomes more complicated. For example, a sedentary person in a cold, wet environment requires enough clothing to restrict heat losses. This clothing would also have to allow escape of normal insensible perspiration. If the body could not rid itself of this vapor, it would collect in our clothing, eventually turn to liquid and then ice, and would lead to rapid hyperthermia. In some cases, then, clothing must serve as both a heat loss barrier and as a screen pervious to moisture.

The rationale behind any clothing system, thin or thick, scanty or complete, is that it allows thermal equilibrium through effective and efficient use of heat transfer mechanisms. We might explore further why consideration of these mechanisms is important in clothing design.

The simple act of touching something may alter the insulation value of clothing by enhancing conduction. Gripping a snow shovel compresses gloves and reduces the amount of air space or insulation. Heat loss through the soles of shoes and boots can be substantial because they are in contact with cold ground or snow. A reversal of the heat problem occurs under high temperatures. Thicker soles with an insulating pad are the answer to heat loss during the winter, and thick, wide soles, but with open tops help in summer. Each protect from heat conduction, but one holds heat in and the other allows for cooling.

Convection can promote fairly rapid heat loss, particularly when advection is added to moving air produced by the body. We discussed this earlier as "wind chill". If clothing is very porous and thin, convection will operate well. Clothing can prevent heat loss by both convection and advection, if it is windproof, such as a nylon shell, and can retain still air, like down or felt.

The tightness of fit is also important in controlling both heat and moisture loss. In hot weather, the flapping of loose clothing can actually be helpful, but under cold conditions, similar movement can reduce the effective insulational value of clothing by fifty percent or more because convection pumps air warmed by your body out and brings cold air in. One usually does not see tight wristbands or anklets around warm season clothing for this reason; in cold conditions, however, such tight closures are essential.

---

2. We are excluding style as a consideration and will not discuss theories on the origin of clothing. For an account of such topics, see Horn (1968) on the reference list.

Radiational heat losses are not as great or rapid as those from other mechanisms. Radiation loss occurs principally from exposed skin, and it is easily stopped by even thin clothing. Protection of such loss from the face, neck, and head is essential during winter. The head and neck comprise some ten percent of total surface area and are the two regions where blood flow is maintained at the same rate at all times because of the brain's oxygen requirements (Table 6.1). Covering the head and neck easily prevents both radiant and convective heat losses. Even at temperatures as high as 15°C, about one-third of actual body heat is lost from the head; for a clothed person without headgear at $-40°C$, the figure is 100 percent.

Table 6.1. Surface Area Percentages of the Human Body

Head	9
Neck	1
Arm	9
Front	18
Back	18
Legs	18

As we have noted several times, clothing also has a tremendous effect on the rate of vapor passage and this in turn strongly influences evaporative cooling. Under hot, wet conditions, little clothing is worn. This aids both direct heat loss as well as evaporative cooling. Conversely, in hot dry climates, the ability to absorb perspiration in clothing is desirable because evaporation from the cloth helps cool the body, yet can retard the rate of moisture loss at the same time. Under cold conditions, either wet or dry, heat loss by evaporation must be dealt with carefully.

Finally, respiratory losses (evaporational cooling) can be substantial also. Under very cold conditions, breathing allows much heat to escape to the atmosphere as well as creating a situation in which air must be heated by the lungs during use. Any device which allows pre-heating, such as a mask, will decrease such heat loss.

Housing

A house is functionally nothing more than a large suit of clothes, rather rigid to be sure, but still constructed to place substances with certain physical properties between humans and their environment. Clothes suffice to protect us individually. The house offers new opportunities. A house can enclose large volumes of space and thus create an environment in which constant stripping or adding of protective clothing layers is unnecessary. A house also permits group living and companionship, as well as protection from animals—including other men.

Important factors affecting a house's ability to provide a protected environment are its size, shape, orientation, materials, openings and setting. If we examine these factors individually and apply them to contemporary as well as traditional dwellings, we can see how these large factors are combined to modify environmental threats.

If we use a "suit of clothes" analogy for housing, then the model posed earlier for human comfort should be applicable. Recall that:

$$\pm B = M \pm (\pm R \pm C \pm E)$$

for the human body. Suppose we relabel B as the net energy balance of the exterior of the house (its skin), M as heat produced not only by humans in metabolism but including such things as cooking and space heating as well. R can be combined with E as sensible (heat detectable by the human body) and latent heat exchange between the inside and outside of the house. C represents conduction between the house and exterior. We can add a term which we will label S for the net energy storage in building materials and the inside of the house. Our model now says that the energy balance of the outside of the house must equal energy processes that produce and exchange energy between the inside and outside. This is merely a summation of the first and second laws of thermodynamics. It also is a simple model that will allow us to state the principles of constructing environmentally appropriate housing in a form that we already are familiar with.

In the simplest model, B is controlled by the diurnal solar cycle. Solar radiation, the major source of energy, strikes a dwelling during the day. It is absorbed and transmitted to the interior by conduction or directly if there are windows or openings; some also is reflected. The amount of energy absorbed at the dwelling surface depends upon many things. The most important factor is the amount of surface the sunlight strikes and its orientation to the solar ray. Also important are the materials comprising the structure.

Some materials such as adobe absorb a significant fraction of incoming radiation, transmit some, and reflect little. This means that a great deal of the insolation energy is converted to heat. Some of the heat is radiated to the air surrounding the house because it is invariably cooler than the sun-heated surfaces. Some of the energy is conducted toward the interior. At night, there is no external energy source. Heat energy now moves by conduction from the interior back toward the exterior and finally by radiation from building surfaces to the outside air. Unlike the human body's situation, the overall net (Incoming-Outgoing) energy balance is negative, because the house is almost always warmer than its surroundings.

In some cases, total M produced by planned and unplanned use of various sources, including that given off by the occupants themselves, is very large. We try to regulate most sources of M because it is the easiest thing that can be controlled to produce a pleasing interior environment. Windows are one example. A warm kitchen can be cooled by simply opening them.

Furnaces are used to do the opposite. Such heat regulation can be viewed as analogous to the function of the primitive brain in trying to control the heat conserving and heat dissipating mechanisms of the body. If we view the house thermostat similarly, then the analogy is even more complete. The thermostat does nothing but sense and respond. If left at one setting, it assumes control without further conscious actions by humans.

R is controlled by the temperature gradient from the building surface to the environment, and the wind speed. Like people, buildings have a thin layer of stationary, cool air around them; heat exchange across this layer occurs by conduction. Since air is a good insulator when calm, heat loss by conduction is slow and minimal even though temperature gradients between the building surface and the air are quite large. As wind speeds increase, turbulence acts to make this thin layer thinner or nonexistent. Heat is then lost by convection, a much more rapid process. Gusty winds also rapidly alter pressure relationships between the exterior and interior of a dwelling which creates a "pumping" action that can "suck" heat out of a house. This is analogous to some action produced in clothing when a person walks or winds collapse and expand different parts of our garments. Air-carrying heat seeps around cracks in the walls, through chimneys and rattling windows. This, too, adds to exterior-directed heat losses.

Wetting dwelling surfaces would keep the interior cooler in the same way that damp clothing lowers skin temperature, but unlike clothes, houses are not easily wetted by artificial means. One can add vegetation to the exterior walls, however, which has the same effect.

C is difficult to describe because the amount of energy loss or gain is dependent on the amount of surface touching the ground and the nature of the contact. It can often be regulated to minimize adverse additions or losses of heat from or to the interior of the dwelling through use of insulation.

Energy exchanges between a person inside the dwelling and the interior are governed by the temperature of the air and wall temperatures. For example, if you are sitting nude in a room with ambient air temperatures in the mid-twenties celsius, then you will likely lose about two-thirds of your body heat by radiation, about twenty percent by evaporation and ten percent by convection. Increasing ths room air temperature to the mid-thirties will cause reapportioning to about 4, 90 and 6, respectively. Activity will cause different values at both room temperatures. The addition of heat to the room, changes in wall temperature, and the addition or deletion of clothing will determine which fractional pattern occurs.

Clothing and Housing as Adjustments to Environments in Different Biomes

Let us now look at how man has used clothing and housing (shelter) to offset specific environmental factors in each of the major biomes. Figure 6.1 will be the framework for our discussion. This figure shows the various temperature-moisture requirements that clothing and housing must meet.

Many of these energy exchange ideas are incorporated into shelter by permanent or transient migrants, and often energy subsidies are applied. For example, elements of Pubelo architecture from the American Southwest are repeated in dwellings in urban areas. Air conditioning by evaporative cooling is added, an environmental modification that takes advantage of certain characteristics of the environment (high temperatures, low humidites), but which requires electricity to operate, a subsidy not available to earlier settlers. In some cases, totally inappropriate shelter adaptations are used in desert areas such as thin-walled, improperly insulated frame houses; the energy subsidy rises rapidly under such circumstances. As we shall see, western cultures have adopted such technological approaches under a variety of environmental circumstances; native populations cannot afford to.

Tropical Rainforest

Minimal clothing and housing are needed in this biome. High humidities hinder body heat loss by evaporation. Radiation, convection and conduction also are minimally effective because air and surface temperatures of surrounding objects often are too high to facilitate much heat loss. Exposing as much of the body as possible, frequent bathing, avoiding unnecessary exertion at all times but particularly at noon, and using shade are behavioral adjustments that prove effective in staying comfortable (Fig. 6.2).

In one housing strategy, maximum energy transfer from interior to exterior is promoted while still providing protection from solar radiation. Walls may be porous or nonexistent (such as thatch or woven rattan) to take advantage of daytime breezes. Air motion also increases evaporative cooling of the body. Thin, porous floors are raised off of the ground facilitating air movement and heat removal by advection. In regions where nighttime temperatures are low, a second strategy is common. A house may be constructed with virtually no openings in order to conserve daytime heat. In both strategies, dwellings are typically grouped in a clearing on a slight rise in the ground surface, if possible. Anthropogenic heat usually is not a problem in either strategy because cooking typically is done away from the dwelling.

Figure 6.2. Tropical rainforest biome clothing.

Monsoon Forest

Because of seasonally heavy rain, permanent dwellings have porous walls and large, shuttered openings (Fig. 6.3). This house form means daytime breezes can be taken advantage of for evaporational cooling, and yet shutters can be closed at night to minimize advective cooling. Radiational cooling to the outside is also decreased by blocking openings. Roofs often are highly angled so that whatever heat is present can rise by convection to the peak and exit out of roof vents or through the roof itself and the heavy rains are rapidly shed to avoid waterlogging or crushing the roof. As in most of the tropics, anthropogenic heat is avoided whenever possible. Cooking is done away from the dwelling. Houses in this biome frequently are raised off the ground on stilts or stone blocks. The raised floor diminishes conduction from the ground to the floor of the dwelling, decreases problems with insects, and allows cooling breezes to strike the floor. Promotion of evaporational cooling is paramount in clothing design and use as well. Clothing practice is similar to that in tropical rainforest regimes except that more of the body is covered to protect from the evaporative cooling accompanying the very wet periods.

Figure 6.3. Monsoon biome housing.

Figure 6.4. Bedouin tent.

Tropical Savanna

Loose, long robes are the most familiar clothing style found in this biome. Direct solar radiation is kept from the body, and perspiration and evaporation (evaporative cooling) are encouraged. The Bedouin tent is a classic housing adaptation to this type of changing regime (Fig. 6.4). It consists of cloth over sticks and is easily portable in keeping with extensive grazing activities. Felt has tremendous resistance to conduction and hot air penetration and keeps solar radiation out. Tents have only one opening to the outside and this is oriented away from the direct rays of the sun. During the more humid part of the year, the sides can be raised to allow passage of breezes to the interior to aid evaporational cooling.

Figure 6.5. Caveform housing in deserts. These particular dwelling units are for priests attending the Diety represented by the statue.

Deserts

Protection from solar radiation and prevention of water loss are the two elements of major concern. All housing and clothing must solve these problems. Houses typically are caveform or actual caves (Fig. 6.5). Walls and roofs are thick, usually brick or some native conglommeration of dried mud and straw (adobe). Houses typically have small windows, which keep direct solar radiation from reaching the interior. Thick walls eliminate hot, dry winds which would increase thermal stress, and allow only slow penetration of heat from the exterior to the interior (Fig. 6.6). In addition such housing can hold heat to offset the often freezing nighttime temperatures that result from cloudless skies. Recall that deserts have strong diurnal temperature variations. Street architecture may also be used to help prevent temperature-radiation

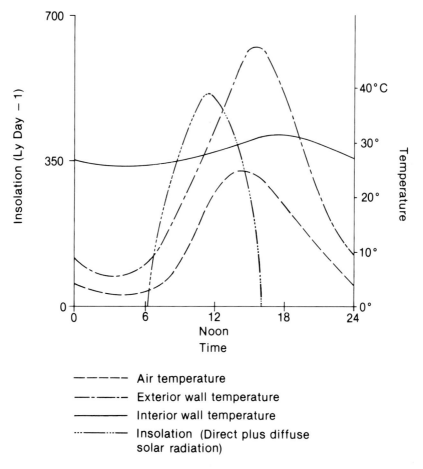

Figure 6.6. Exterior and interior conditioned for a adobe dwelling.

stresses in hot, dry regions. In small towns and cities of the desert biomes, dwellings often are narrow, tall structures (Fig. 6.7). Both characteristics keep solar radiation penetration at the street level at a minimum.

Clothing, like housing, must permit evaporative cooling while protecting the body from incident radiation. Long robes and a head covering are favored by many native populations. The open shirt and short pants seen in resorts can only work when there is abundant water and air conditioning is available. Otherwise, water loss alone would kill the wearer in a short time.

Prairie/Steppe Biome

As in all temperate biomes, clothing patterns are seasonal. Warm clothing is necessary to meet the rigors of winter; less clothing is needed for the hot summers. In winter, clothing must protect from radiation losses while

Figure 6.7. Street scene in a desert biome.

Figure 6.8. Sod house on the Great Plains.

permitting moisture escape. In summer, the need is to enhance evaporation losses. Housing also must meet changable conditions. Early housing adptations in the grassland biomes of the United States included sod houses which were warm in winter, cool in summer (Fig. 6.8). As the years passed and energy to heat became cheaper, sod houses were replaced by frame houses, a form not naturally adaptable to a prairie/steppe environment. Today, with expensive energy, interest is again focused on earth-sheltered homes as a way of cutting energy costs (Fig. 6.9).

Temperate Evergreen Forests

As in the Prairie and desert areas, protection from high radiation loads, and rapid water loss is a necessity during the summer. Thick-walled houses with many of the characteristics found in Desert biomes admirably meet these needs. Humidities are low during the warm season, and fountains or other sources of evaporating water are often used to provide a more comfortable atmosphere. The open courtyard in the center of the home, with the walls giving privacy from neighbors, is common. One thinks here of the Roman

Figure 6.9. Modern earth-sheltered house.

townhouse or the Spanish ranch house as a model (Fig. 6.10). Cool winters also occur in these biomes. Heat must be retained at such times or the body, even inside the house, would cool rapidly, losing heat to the walls of the dwelling.

Temperate Deciduous Forests

In such areas, houses with small windows, thick or heavily insulated walls often must be supplemented by the addition of heat for nine months of the year especially in the colder, poleward sections of this biome. Any sunshine can be taken advantage of because excessive radiation is not a problem. Similarly, clothing must permit heat retention during the winter while allowing for moisture dissipation. Summer clothing must allow evaporation for body cooling.

In the warmer, southern portions of the biome, houses typically combine tropical housing features, such as wide porches, shading, and large openable windows, to dissipate summer heat while allowing entry of cooling breezes.

Figure 6.10. Courtyard and central fountain within a Spanish ranchhouse once typical in temperate evergreen forest biome regions.

Figure 6.11. Housing in the warmer portions of the temperate deciduous biome.

High ceilings on the interior and an open plan to the room arrangement also help. The cool season may see the addition of shutters and a sealing up of the house against the winter cool (Fig. 6.11).

Temperate Rainforest

Homes must retain heat during the longer, cooler, wetter portion of the year, and dissipate it during the shorter, warmer, somewhat less humid times. Evaporational cooling through clothing is not a prime consideration for the most part given the cool mild nature of the climate throughout most of the year, rather heat retention is necessary.

Boreal Forest

Clothes must be designed to permit maximum heat retention during the cold season; radiational, conductive, and convectional losses are kept to a minimum. Porosity to moisture is also important in order to rid the body of vapor. Several layers are favored to permit shedding when necessary. There is little direct solar radiation during the winter, but a great deal during the short summer. Housing requires heavily insulated exterior surfaces because it is the cold season which clearly predominates. Window openings must be

Figure 6.12. Housing in a boreal forest region. (Courtesy of W. Johnson.)

very small to cut radiational losses and are shuttered at night even in the warmer months of the year (Fig. 6.12). In many areas, insulated floors are a necessity in order to keep anthropogenic heat from being conducted to the soil surface and literally melting the ground. Melting creates an unstable surface and an infirm foundation.

Tundra

The problem is a simple one: how to keep warm in an environment devoid of natural heat sources. Heavy insulation is the rule, and innovative design, such as the igloo, often provides conditions necessary for human comfort. Igloos are exceptionally good protective devices against harsh conditions. Inside temperatures are not high but within the normal limits tolerable by humans (Fig. 6.13). Today, wooden shacks are being built by natives with assistance from government sources; these houses are not adapted thermally to Tundra regimes (Fig. 6.14). They require tremendous amounts of energy in order to keep them comfortable. A great deal of non-native clothing using modern fibers and insulation techniques are being utilized. The principle of layered clothing, small air spaces, a covered head and extremities, and respiratory preheating still hold true, however, even with modern designs. An example of a traditional design is shown on Figure 6.15.

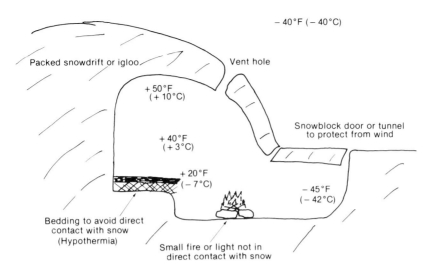

Figure 6.13. Temperature conditions on the inside and outside of a snow cave dwelling.

Figure 6.14. Government subsidized housing on the tundra. (Courtesy of W. Johnson.)

Figure 6.15. Eskimo outerwear.

Increasing elevation in any biome would produce a decrease in heat and moisture, and decreases in daily as well as annual temperatures. Protection from high radiation loads is essential, but not as critical as in hot dry areas. Radiational cooling assumes greater importance with elevation due to thinning of the atmosphere and thereby less ability to return some of the lost radiation back to the earth's surface and lower atmosphere. Houses are designed to keep heat at low levels within the dwelling, and roofs must resist large heat losses to the night sky. Folk architectural practices in Europe often dictated an organic roof, such as sod, to meet the heat loss requirement. Many early dwellings in this country located in the Rockies were small and used heavy timbering for roofs as a heat saving measure.

References

AIA Research Corporation. *Regional Guidelines for Building Passive Energy Conserving Homes.* U.S. HUD. #PDR-355. Washington, D.C.: GPO, 1978.

Horn, M. J. *The Second Skin: An Interdisciplinary Study of Clothing.* New York: Houghton Mifflin Co., 1968.

Oke, T. *Boundary Layer Climates.* New York: John Wiley, 1978.

Oliver, J. *Climate and Man's Environment.* New York: John Wiley, 1973.

VII
Closing Thoughts

All natural environments are composed of a great variety of factors which change from one place to another. Unlike animals and plants, which are adapted to a limited range of environmental factors, humans can use or modify environments and behaviors in ways that will support life and comfort. The scope of coping strategies depends upon the amount of available energy subsidies.

In the past, man overcame environmental limits by adjusting to natural constraints. He used the environment itself to support human life by living in cliffs or pits, by facing his housing toward the equator so that it would benefit from solar energy or by migrating to follow supplies of water, food or seasons of warmth. More recently, man has used technology and fossil fuels to overcome the energy limits imposed by natural biomes. He has approached the environment more as an enemy, its constraints as threats, and himself as engaged in a "war against nature."

How will man cope with the natural world in the future? Human ingenuity continually devises new methods of coping with environmental limits. This process will never stop, for, if the natural environment seems inexhaustively flexible, so is human ingenuity.

It is our hope that by reading this book, you have developed a broader appreciation of some of the major ingredients in the man/land equation.

Index

Acute Mountain Sickness (A.M.S.), 43, 44
Adderley, E. E., 26
Aggradation, 21, 24
Agricultural Cycles (see also Cycles)
  Disease and Insects, 13
  Droughts, 15
  Growing, 11
  Phenological Events, 12, 13
A.I.A. Research Corp., 109
Allen's Rule, 44, 46
Altruism, 50, 58
Annual, 12, 24
Ardrey, R., 59
Armitage, R., 26
Aschoff's Rule, 24, 26
Astronomical Cycles, 13
Atmospheric Cycles, 15
Autonomic Nervous System 49, 58

Baker, P. T., 48
Basal Metabolism Rate, 36, 46
Bates, M., 8
Beets, J. L., 26
Bensinger, T. H., 48
Bergmann's Rule, 44, 46
Biennial, 13, 24
Billings, W. D., 88
Biological Cycles (see also Cycles), 11, 18, 24
Biological Limits, 2, 29, 66
Biomass, 64, 68
Biome, 46, 61, 89
  Carrying Capacity, 65, 86
  Clothing for, 89, 91, 95
  Energy Relationships, 91
  Housing for, 92
  Productivity of, 63, 86
  (Also see Arctic, Cold, Desert, Temperate and Tropical)
Biorhythms, 18, 24
Biosphere, 15, 24
Boaz, F., 2, 8
Body
  Size and Shape, 44
Bowen, E., 26
Bradley, D. M., 26

Brain
  Evolution of, 51
  Limbic, 49, 58
  Neocortex, 49, 58
  R-Complex, 49, 58
  Related to Human Needs, 52
  Right vs. Left Hemisphere, 50, 51
Brezowsky, H., 26
Brier, G., 26
Brown, F. A., 26

Calories, 37, 46
Campbell, D., 26
Cannon, W. B., 48
Carnivore, 63, 86
Carrying Capacity (also see various Biomes), 65, 86
Chappell, T., 48
Chronobiology, 18, 24
Clapham, W., 88
Clements, F., 48
Climate (also see various Biomes), 15, 86
Clothing (also see various Biomes), 90
  Design, 90, 91
Cloudsley-Thompson, J., 26
Cold Biomes, 82
  Alpine, 84
  Arctic, 84, 106
  Boreal, 83, 105
Colquhoun, W., 26
Crop Calendar, 12, 24
Cycleology, 21, 24
Cycles, 9, 24
  Adaptive Role, 15
  Agricultural (see also various Biomes), 10, 11
  Aschoff's Rule and, 18
  Astronomical, 11, 13
  Atmospheric (also see various Biomes), 11, 15
  Behavior and, 32, 55
  Biological, 11, 18, 24
  Circadian, 20, 24
  Cultural, 11, 21
  Disease and, 13
  Endogenous, 18, 19
  Exogenous, 18

113

Frequency of, 9
Geological, 21
Holidays as Markers, 14
Light and Dark, 18
Lunar, 15
Monuments Marking, 14
Periodicy, 9
Protective Aspects of, 10
Solar, 14
Table of Selected, 11
Temperature as a Regulator of, 11, 19, 20

Decision Making Process, 52, 53
Decomposer, 63, 86
Degradation, 22, 24
Dehydration, 28, 42
Desert Biome, 85, 99
Diastrophism, 24
Diet, 32, 33
    Daily Requirements, 31, 32
    Historical Changes in, 34, 35
    Proteins, Fats, Carbohydrates, 34
Disease,
    Humans and Plants, 13
Diurnal, 10, 24
Downs, R., 59
Dubos, R., 48

Ecological Anthropology, 3, 7
Ecotone, 27, 46
Edema, 44, 46
Edman, M., 48
Electrolytic Balance, 29, 32
Endogenous, 18, 19, 24
Endorphine, 31, 46
Environmental Determinism, 1, 7
Equinox, 24
Exogenous, 14, 18, 24

Feedback Mechanisms, 61
Fertilizer, 73, 86
Fisher, K., 48
Folk, E., 39
Food
    Crops (see Agriculture and various Biomes)
Food Chain (see also various Biomes), 63, 86
Food Web (see also various Biomes), 63, 86
Frank, J. D., 48
Frequency (see also Cycles), 9, 24
Fuqua, M., 48

Genetic
    Adaptation to Altitude, 44
    Role in Decision-Making Process, 52–55
Geological Cycles, 11
    Aggradation, 21, 24
    Cycleology, 21, 24
    Degradation, 22, 24

Goldenweiser, A., 3, 8
Grassland, 68, 69, 70, 76, 81
Greeks, 1
Greenhouse Effect, 17, 25
Grey, W., 48
Growing Season (see various Biomes), 10, 25

Hackett, P., 48
Halberg, F., 20, 26
Hall, E., 59
Hardy, J., 48
Heat Stroke, 41
Herbivore, 63, 86
Hibernation, 10
Hierarchial Differentiation, 61, 86
Hilts, P., 26
Holden, C., 59
Holidays
    As Cycle Markers, 14
Homeostasis, 27, 36, 46
Homeostatic Mechanisms, 36, 38, 39
    And Maintenance of Temperature, 39
    BMR (Basal Metabolic Rate), 36
    Biochemical Constraints, 29
    Electrolytic Balance, 29, 32
    Heat Gains and Losses, 36
    Heat Stroke, 41
    Hormones, 31
    Hyperthermia, 39, 46
    Hypothermia, 39, 46
    Hypoxia, 46
    Moving Air, 40
    Neutral Zone, 29
    Nutrients, 30
    Responses to Thermal Stress, 38, 39
    Role of Diet, 32
    Temperature Effect on, 36
    Water Needs in, 28
Hormones, 31
Horn, M., 91, 109
Housing (see also various Biomes), 92
    As Adaptive Process, 95
Human Determinism, 4, 7
Huntington, E., 8
Hyperthermia, 39, 41, 46
Hypothermia, 39, 40, 46
Hypoxia, 43, 46

Isolation, 61, 86

Klein, R., 26

Latent Heat of Evaporation, 37
Laterized, 73, 86
Light, 18
Limbic System, 49, 58
Loye, D., 59
Luce, C., 26

Macrospace, 57, 58
Mesospace, 57, 58
Metabolism, 36
 Basal, 36
 Typical Values for, 36, 37
Microspace, 56, 58
Mills, J., 26
Minimum Daily Requirements, 31, 33
Mironovitch, V., 26
Mitchell, H., 48
Model, 55, 58
Monotheism, 1
Morphological Adjustments
 to Altitude, 44
 to Temperature, 44

Needs
 Pyramid of, 52
Negative Feedback, 62, 86
Neighborhood, 57
 see Space
Neocortex, 49, 58
Neutral Zone, 29
Nocturnal, 10, 25
Nutrients
 Deficiencies, 30

Oke, T., 109
Oliver, J., 109
Optimum Conditions, 27, 47
Oxygen, 43
Oyle, I., 48

Palmer, J., 26
Peptide, 47
Perennial, 13, 25
Periodicity, 9, 25
Phenological Events, 12, 25
Picardi, G., 48
Portrous, J., 59
Positive Feedback, 62, 86
Possibilism
 Role of Technology in, 2, 7
Prairie Biome, 81
Predation, 62, 86
Probabilism,
 Role of Culture in, 3, 7
Producer, 63, 86
Productivity, 63, 86
Proteins, 33

Radiation (see various Biomes)
 Solar, 61, 63
Raloff, J., 26
Range of Tolerance, 27, 46
 Light,
 Temperature,
  Maximum, 41
  Minimum, 40
  Optimum, 29
 Water, 28, 42
Rappoport, R., 8
R-Complex, 49, 53
Relief, 21, 25
Respiration, 63, 87
Revolution, 14, 25
Rotation, 14, 25

Saarinen, T., 59
Sagan, C., 48
Sandlow, S., 26
Saunders, D., 26
Scale and Behavior (see also Territoriality)
Seasons (see also various Biomes), 14, 25
Self-Regulating Mechanisms (see also
  Feedback and Homeostasis), 61,
  87
Shelford, V., 88
Slager, U. T., 33
Sleep
 as a Cyclic Phenomenon, 10
 Hibernation, 10
 Theories for, 10
Soils (see various Biomes)
Solstice, 14, 25
Sommer, R., 59
Stea, D., 59
Steppe Biome, 70, 81
Steward, J., 8
Still, H., 26
Strategies
 Coping, 52, 64
 Energy Subsidized, 66
 Learning, 67
Stresses
 Thermal, 40, 41
 Water, 42
Stroke
 Heat, 41
Structure
 of Biomes, 61
Sunspot, 9
 Cycles, 25
Surface Area
 to Volume Ratio, 45
Symbiosis, 6, 7
Synchronization
 Light, 18
 Temperature, 18
Synodic Time, 15, 25

Technology
 as Adaptive Mechanism, 66
 Range of Cycles, 53
Temperate Biomes, 77
 Deciduous, 81, 103
 Evergreen Woodland, 78, 102

115

Prairie and Steppe, 81, 100
Rainforest, 77, 105
Temperature
  and Water Needs, 42
  as Weather Element, 62
  as Zeitgeber, 18
  Effects on Homeostasis, 18
  Neutral Zone, 29
  of Body Core, 29
  Role in Hyperthermia, 41
  Role in Hypothermia, 40
  Role of Clothing, 91
  Thermal Equilibrium, 36, 38, 39
  Wind Chill, 40, 41
Territoriality, 49, 56, 58
  Advantages of, 56
  Size of Space, 56
Thoracic Cavity, 40, 46
Triune Brain
  see Brain
Tromp, S. W., 48
Trophic Levels, 63, 87

Tropical Biomes, 73
  Monsoon, 73, 97
  Rainforest, 73, 95
  Savanna, 76, 98
Turner, F., 21

Uni-directional Energy Flow, 61, 87

Vitamins,
  see Nutrients

Water Needs, 42
Watt, D., 88
Weather Elements, 62, 87
  see also various Biomes
Weiner, J., 48
Wilson, E., 48
Wind Chill, 40
Woodbury, M., 26

Zeitgeber, 18, 25